CHANGE!

Isaac Asimov

CHANGE!

Seventy-one Glimpses of the Future

BOSTON

Houghton Mifflin Company

1981

Library of Congress Cataloging in Publication Data
Asimov, Isaac, date
Change! : seventy-one glimpses of the future.

1. Science — Popular works. I. Title.
Q162.A77   500   81-4235
ISBN 0-395-31545-X   AACR2

Printed in the United States of America

P   10   9   8   7   6   5   4   3   2   1

These articles originally appeared in *American Way* and
are reprinted by permission. Copyright © 1974, 1975,
1976, 1977, 1978, 1979, 1980, 1981 by American Airlines.

To Walter Damtoft
with whom it is a pleasure to work

CONTENTS

CONTENTS

Contents

INTRODUCTION

BACK IN 1974, it occurred to the good people at *American Way Magazine* (the inflight magazine of American Airlines) that it would be nice to have a monthly column on some aspect or other of the future — a column they would call "Change."

When they asked me to assume the responsibility for the column I said, sadly, that I would love the job, but I had an irrational fear of flying and never flew in airplanes. What's more, this was well known, since I had frequently been laughed at in print over the matter, even in such usually grave periodicals as the *New York Times*.

They smiled kindly and said that it did not matter, so I seized the chance to do the column with both hands.

I admit that, at the start, I feared I might not be able to think up different aspects of the future for long, but human life and society have grown so complicated that more than six years later I have not exhausted the subject. I hope the column will continue indefinitely.

I must warn you that none of the futures described in this book are inevitable, and some of them are, perhaps, incompatible with others. But then, that should be no surprise. I do not predict the inevitable, but the possible, the conceivable — sometimes only a dream.

The actual future will be what circumstance, the human will, and the human intelligence make it, and we can only hope that all conspire to produce something good. My role, as a futurist, is to reconnoiter the territory up ahead so that humanity, in its travels through time, may have a better notion of what to aim for and what to avoid.

You may notice aspects of the future I *don't* discuss in the book. Well, give me time. The column continues, and perhaps someday there will be a book entitled *More Change!*

ISAAC ASIMOV
July 1980

CHANGE!

1 THE FUTURE OF FUTURISM?

In 1941 I wrote the first in a series of stories that evolved into *The Foundation Trilogy.*

In it I invented a science I called "psychohistory" — the statistical study of every kind of sociological trend, using mathematical techniques advanced far beyond the present, for the story was set many thousands of years in the future. As I told it then, psychohistory could predict, with a high degree of accuracy, the general social movement that would take place in the future, even though the actions of individual human beings remained unpredictable.

I did not make up the concept out of nothing at all. I had in mind what is called the kinetic theory of gases: A sample of gas is made up of trillions of trillions of molecules, each one of which moves about in random directions, at random speeds, and bounces off its neighbors in random ways.

If we analyze this random movement mathematically, however, we end with an almost absolute determinism. We know exactly what will happen to the gas if we compress, expand, heat, or cool it. The random behavior of the individual molecules makes the gas as a whole a completely predictable system.

With this in mind, I explained in *The Foundation Trilogy* that psychohistory would work with reasonable exactness only if the actions of many, many human beings were being analyzed, so that individual unpredictability would melt into the common crowd and be lost. In this way, personal free will (random movement, if you like) would be maintained, but the overall direction of change would become known.

1

Second, psychohistory would work only if the general population did not know what the predictions were, so that their behavior would remain truly random.

The psychohistory in my *Trilogy* was played out against the vast background of a Galactic Empire of millions of worlds and trillions of human beings; of the fall of that Empire and the Dark Age that followed. Being young and eager, I managed to infuse into those stories enough plausibility to catch the hopeful dreams of my readers. As a result, I received many letters from those who were convinced I had done research in the field of psychohistory and wanted more detail.

I had to write to each of them and explain that there was no such science, and that there probably might never be such a science. I had to explain that I had simply made it all up back in 1941.

Then, some time ago, I had a conversation with my nephew, Daniel Asimov, a professor of mathematics. His field of study has been topology — the study of those properties of geometrical figures that don't change if the figures are deformed. It is a fascinating field, but he is eager to find something that might prove to be even more fascinating.

He was thinking, he said, of tackling the mathematics of society: studying its properties, its movements, its changes, and so on.

"Can that be done?" I asked in astonishment.

Danny shrugged. "We can make a start, perhaps."

I said, "But that's psychohistory."

"I know," he said.

Is it possible that the future of futurism is, after all, psychohistory? Will the time really come (as I had casually dreamed it would in 1941) when it will be possible to understand what makes humanity in large groups behave the way it does? Will it be possible to see the future toward which society is trending? Will it be possible to understand what alternatives exist; what changes must be made to bring about another, a more desirable, future; and how to achieve it most efficiently and rapidly and at least cost?

If my instinct is correct, would it mean that all this alteration of the direction and movement of society in search of a more efficient pursuit of happiness would be carried out without affecting individual free will?

Is this elitism, though? Is it wrong and paternalistic to try to design a future?

If it is, we are all guilty of it anyway. All governments try to direct the future. All generals do. All corporation heads do. All parents do. All people do. *You* do. No one *really* wants to stumble about blindly.

The trouble is that right now we know so little about what makes people and society tick that all of us, from individuals to governments, are forever groping blindly, even though we don't want to. It may be that, with psychohistory, the time will come when we will do exactly what we are trying to do now — produce a desirable future — but do it successfully!

2 THE INDESTRUCTIBLE

SOME of the most impressive changes of the last century have involved the vehicles for the entertainment of human beings.

We went from player pianos to record players; from vaudeville to motion pictures; from radio to television. We added sound to movies, images to radio, color to both. And there seems no doubt we can go further still.

By using laser beams and holography, we can produce

three-dimensional images more detailed than anything that can be obtained by ordinary photography on a flat surface. By using modern taping procedures we can produce TV cassettes on any subject, so that an individual can play what one wants on one's own set at a time to suit one's convenience.

Every new advance outmodes the older devices as people flock to the technique that gives them more. The motion picture killed vaudeville, television killed radio, and color killed black and white. Three dimensions will no doubt kill flatness, and cassettes may kill mass-produced, general-purpose television.

What is it all tending toward? What will the ultimate be?

I watched a demonstration of TV cassettes once, and I couldn't help noticing the bulky and expensive auxiliary equipment required to decode the tape, put sound through a speaker and images on a screen. Surely the direction for improvement will be miniaturization and sophistication, the same processes that in recent decades have given us smaller and more compact radios, cameras, computers, and satellites.

We can expect the auxiliary equipment to shrink and, eventually, to disappear. The cassette will not only become self-contained, but will hold the tape and all the mechanisms for producing sound and image.

With miniaturization, the cassette should become more and more nearly portable; eventually light enough, perhaps, to be carried under one's arm. It should also require less and less energy to operate, reaching the ultimate ideal of requiring no energy at all.

An ordinary cassette makes sounds and casts light. That is its purpose, of course, but must sound and light obtrude on others who are not involved or interested? The ideal cassette would be visible and audible only to the person using it.

Present-day cassettes involve controls, of course. They must have an on-off knob or switch and devices to regulate color, volume, brightness, contrast, and all that sort of thing. Naturally, the direction of change will be toward simplification of controls. Eventually, they will have a single knob — or perhaps none at all.

We could imagine a cassette that is always in perfect adjustment; that starts automatically when you look at it; that stops automatically when you stop looking at it; that can play forward or backward, quickly or slowly, by skips or with repetitions, entirely at your pleasure.

Surely that's the ultimate dream device — a cassette that deals with any of an infinite number of subjects, fictional or nonfictional, that is self-contained, portable, non-energy-consuming, perfectly private, and largely under the control of the will.

Must this remain only a dream? Can we expect to have such a cassette some day?

The answer is a definite yes! Not only will we have such a cassette some day, we have it now; not only do we have it now, we have had it for many centuries. The ideal I have described is the printed word — the magazine, the book, the object you now hold — light, private, and manipulable at will.

Do you think that the book, unlike the cassette I have been describing, does not produce sound and images? It certainly does.

You cannot read without hearing the words in your mind and seeing the images to which they give rise. In fact, because they are *your* sounds and images, not those invented for you by others, they are better.

All amusement media other than the printed word present you with prepackaged images, or sound, or both, in greater and greater detail as technology improves. The result is that

5

the media demand less and less of you. Even musical cues and laugh tracks are provided to draw particular emotions from you without effort on your part. If reading is difficult for a person (and it is for most) it is to this prepackaging he or she will turn, and a passive spectator he or she will be.

The printed word presents minimum information, however. Everything but that minimum must be provided by the reader — the intonation of words, the expressions on faces, the actions, the scenery, the background must all be drawn out of that long line of black-on-white symbols. The book is a shared endeavor between the writer and the reader as no other form of communication can be.

If you are, then, of that small and fortunate minority for whom reading is easy and pleasurable, the book, in all its manifestations, is irreplaceable and indestructible, for it demands that you participate. However pleasurable spectatorship may be, participation is better.

3 SPEAKING TO ALL THE WORLD

THE WORLD has been getting effectively smaller for a long, long time; now it will fit into almost anyone's backyard. No place on Earth is more than hours away from any other place by fast plane, or more than fractions of a second away by radio and television.

What this means is that any two people on Earth — *any* two — might potentially find themselves having to communicate with each other. What language will they use? If we were to pick any two people on Earth at random, the chances are they'll have to use sign language.

More people on Earth speak Mandarin Chinese than any other language — perhaps 470 million. Almost all those millions are concentrated in China itself, however, and at that they make up only 11 percent of the Earth's population, and perhaps not more than three fifths of the population of China itself.

The next most common language is English, which is spoken by perhaps 340 million people, representing only a little over 8 percent of the Earth's population. But it is a very widespread language, thanks to what was once the British Empire. English is spoken by 10 percent or more of the population in thirty-two different nations and, in every case, by a much larger percentage of the educated and technologically trained in those nations (where it is not already the native tongue).

English is the almost universal language of science, business, and international politics. It might be thought, then, that as the world is knit together more closely by ever-improving means of transportation and communication, English will become ever more dominant and will, in effect, become the global language — either the first or second language of every person on the planet.

But there's a catch!

There could be nationalistic reactions against English. It might well seem to billions of people on Earth that those for whom English is the native language would have an advantage over those for whom it is a learned language; that the English-language heritage in literature and in culture generally would cast all others into the shade.

It is easy to imagine a linguistic revolt: a refusal to speak English or pretending not to understand it. Many French Canadians, for instance, take pride in not understanding English. There might well be movements to make Gaelic and Welsh stronger within the British Isles themselves.

Other languages suffer revolts. India must continue to use English as an official language since the domination of no one Indian language will be permitted by those who speak others. In the Caucasus, Georgians mounted a rare public demonstration against Soviet policy when they protested an attempt to make Russian the official language of their Soviet Republic.

What's the alternative? Interpreters? Whether interpretation is human or computerized, could we trust it? How incredibly easy it would be to make small errors in interpretation and how costly those might be.

Can we have a global language other than English? Which? Surely any language other than English would create even louder objections the world over.

There are artificial languages, of course. The best known is Esperanto, invented in 1887. It is a very sensible language, very easy to learn, but it is essentially a distillation of the Romance languages and might therefore rouse non-European hostility. Besides, artificial languages seem to lack vitality. After nearly a century there are only about 100,000 speakers of Esperanto. Other artificial languages are even less successful.

Yet the problem may well take care of itself, for on a smaller scale, solutions have arisen with no one's purposeful interference. In the Mediterranean world in late medieval times, a "lingua franca" ("language of the Franks," i.e., Europeans) developed among the merchants of the area to take care of absolutely necessary communication. It was a mixture of Italian, French, Spanish, Greek, and Arabic. Similarly, in eastern Asia, various mixtures called Pidgin English were developed for communication across the languages.

As communications around the world improve, and as more and more common folk may want to talk — not just educated businessmen and scientists — "lingua terra," a "language of the world," may arise little by little. English will make up a large part of it, yes; but every other language will add vocabulary, idiom, and grammar.

It could end as a fearful construction, with rules all its own bearing no too-clear mark of any one national language. All human beings would have to learn it in addition to their own languages, and none would have an advantage over others by the mere accident of place of birth. Lingua terra could end with a vocabulary, a flexibility, a richness surpassing any other, so that it could develop a mighty literature of its own.

It might then, by its mere existence, do more to emphasize the familyhood of humanity than a million sermons on the subject laid end to end.

4 THE FALLING BIRTHRATE

FOR AS LONG as our records trace back into the human past, women have always seemed to be subordinate to men. Girls have never had the educational opportunities of boys, nor women the social or sexual advantages of men. The "double standard" has always worked in man's favor, and the important work of the world — in government, in industry, in religion, in science — was almost entirely assumed by men.

There have been exceptions. Some societies have oppressed women less than others, and some women have managed to achieve positions of prominence even in oppressive societies. These *are* exceptions, though. On the whole, women the world over, and through all the ages, have walked a submissive three paces behind their menfolk.

9

CHANGE!

This has been true for so long that not only do almost all men accept the situation as normal and natural, but so do almost all women. Women are the largest of all minorities (they are actually a slight majority) and yet even at this time when it has become unfashionable and downright unrespectable to downgrade minorities, we can still shrug off women's role in the world. It is still possible for men to say in a condescending way that "woman's place is in the home" or "in the kitchen" or "in the nursery."

Of course a vocal and ardent Women's Liberation movement has become a respectable part of the American scene. It makes headlines, but it is viewed with suspicion by most men, and is a source of uneasiness even for most women. Can the feminist movement win out?

Yes, if something happens to make the idea of equality for women desirable to everyone. In that case, it can't fail to win.

Consider the birthrate. It has dropped steadily in the United States over the last decade until, at this moment, an average family includes slightly fewer than two children. This doesn't mean that the American population is dropping; there are still lots of youngsters, born in previous higher birthrate years, who are growing up and will soon have children of their own. It does mean, though, that if the birthrate stabilizes at its present figure, the United States will reach a population plateau of 300 million about seventy years from now.

As the decades go on, however, the population will rise precipitously in the rest of the world and the crunch in food, energy, and other resources will grow rapidly worse. The crunch will become disastrous here in the United States as well as elsewhere, for though our population rise will be smaller than that in the rest of the world, our accustomed per capita use of food, energy, and other resources is much higher.

It will therefore become steadily more important to mankind to *keep* the birthrate low where it is already low, and *make* it low where it is not. We will have to bring the world population down, and while the rise in starvation and disease and violence will help achieve that, they will also shake the fragile underpinning of our civilization. The less we rely on a higher deathrate, and the more we rely on a lower birthrate, the more likely it will be that civilization will survive.

We have every reason, therefore, to suppose that the present low birthrate in the United States will continue — that every effort will be made to force it to continue. It is almost inevitable that each woman will be restricted (by social pressure, at the very least) to having no more than two children, and that women will be encouraged to have only one child, or even none at all, if they so choose.

In that case, what do we do with women? How do we insist that woman's place is in the home or in the kitchen, if the small-family rule results in not much work in the home and kitchen? How, in particular, can we insist that woman's place is in the nursery when the world is going to be very anxious to keep women *out* of the nursery?

It won't do to have women sit around twiddling their thumbs. If we continue to deny them a useful world role in industry, science, religion, and government, what choice will they have but to try to return to the only kind of importance they have always been permitted, nay, urged to assume — their role as mothers?

The aims of feminism will therefore no longer be the property of activists alone — it has already grown respectable and will have to become the settled and serious policy of society in general. The vast change in attitude will be brought about not by the angry or subtle persuasions of speeches, resolutions, or books, but as the inevitable consequence of the falling birthrate that the world will experience and want to continue to experience.

Naturally, such a vast and rapid turnabout will be hard to accept, and it may be that the firmest and most stubborn holdouts against the notions of feminism will be women rather than men. It may be that political, social, and economic equality will have to be forced on many women against their will. We might hope, though, that it will be done as humanely as possible.

NOTE: This essay, which appeared in 1974 as the first in the series, seems to have hit the target. In the years since its publication, the women's movement has continued to make strides the world over and the birthrate has indeed been falling. It has not fallen enough, to be sure, and world population is still going up distressingly, but at least the change is in the right direction.

5 ADULT EDUCATION

ONE CHANGE has been creeping up on us steadily, though perhaps not very noticeably.

In the United States in 1900, there were 3.1 million people over sixty-five out of a total population of 77 million, so that the elderly made up just about 4 percent of the whole. In 1940, it was 9.0 million out of 134 million, or 6.7 percent. In 1970, it was 20.2 million out of 208 million, or nearly 10 percent. By 2000, it is likely to be 29 million out of 240 million, or 12 percent.

Our senior citizens are not only increasing in numbers; they are steadily increasing in terms of percentage of the population. Naturally so, since, in the United States, the average length of life has been increasing and the birthrate decreasing throughout the twentieth century.

Population pressures may force a steady decrease in birthrate the world over in the decades to come, and the continuing advance in science and medicine may increase the average length of life. If so, what has been true of the United States and some of the other industrial nations may become a world phenomenon. All over the world, a larger and larger percentage of the population will be middle-aged, elderly, old.

The most interesting point about such a situation is that humankind as a whole will be faced with an age pattern it has never experienced before. Never! Throughout the world's history, till now, the situation everywhere has been one of short lives and a high birthrate. Those who managed to pass their fortieth birthday were a small minority; those over sixty-five an actual rarity.

Many of our social customs reflect this past. We tend to ignore our oldsters because, in the societies of the past, they were too few to worry about. We tend to assume that they are feeble, dull, and worthless because in the past those few who did manage to survive into middle age were worn out by repeated bouts of infectious disease, by poor diets, bad teeth, hormonal imbalance, and lifelong overwork.

So now, when the aging are stronger, more vigorous, and more alert than they once would have been, we still retire them ruthlessly to get them out of the way, leaving them to wait out life's ending on a park bench.

The easy excuse for this is that the aging are stodgy and conservative, no longer capable of creative thinking. Only the young, we suppose, are fit to blaze new and innovative trails.

Might we not be reasoning in a circle? After all, we only educate the young. A cut-off date is taken for granted in education. We go to school only for as long as it takes us to learn a trade or profession, or to gain the ability to converse in a cultivated fashion, and then we quit — having become *educated* (past tense).

CHANGE!

The notion that one has already been *educated* (past tense) offers one social permission never to read a book again, never to learn anything more, never to have another thought. Is it surprising that, under such conditions, so many people as they grow older become incapable of reading, learning, or thinking altogether?

In a long-life, low-birthrate world of the future, however, this will have to change. As aging people make up larger and larger percentages of the population, it will become more and more dangerous to permit them to remain a nonthinking deadweight on society.

We will have to alter our attitude toward education. Society will have to take it for granted that learning is a lifelong privilege, that education is not a task to be completed, but a process to be continued.

Men and women, growing up and seeing this attitude all about them, will accept as natural the fact that throughout life they will be able to use educational institutions and techniques to learn new things and enter new fields. Under such conditions, accustomed to lifelong learning, why shouldn't they remain creative and innovative to very nearly the end of their lives?

This is a difficult notion to accept. It may seem self-evident that the mental workings of aging individuals simply must become rusty and creaky. That kind of "self-evident" proposition is, however, only a guess based on youth-centered prejudice.

In actual fact, we have never, in the world's history, had a real process of adult education, and we have never given the aging mind a chance to show what it can really do — so we just don't know what it can do.

Given the kind of world we may be entering, we will *have* to give the aging a chance, and my guess is that we will be in for a pleasant surprise.

6 WHO NEEDS MONEY?

MANY PEOPLE consider *computers* to be a dirty word. These gadgets appear to them to threaten dehumanization, to be devices incapable of taking a human look at problems, capable only of doing everything by the numbers.

But then that is also true of the slide rule, the balance, the yardstick. It is true of any device used by humanity to solve its problems mechanically.

Some architect of prehistoric times may have complained bitterly about building structures with the aid of the new-fangled yardstick. "You don't rely on a dead piece of wood to tell you how long to make the stone pillars," he might have grumbled. "You must use the trained eye of a skilled architect, or we will all be dehumanized."

Computers, to be sure, are enormously more complicated than any problem-solving device we have ever had before, enormously faster, enormously more capable of dealing with many factors simultaneously. That's *good*, because never before in history have we had problems so numerous and so complex. We have never had so many people, using so many resources in so many ways, to build up so intricate a society.

We have now reached such a pitch of social and technological complexity that to abandon the problem solving of computers would force us back to a stage we were at thirty years ago, and we could not feed the billion and a half people we have added to the world's population in those thirty years. Airlines, banks, almost any industry, and certainly the government, could not fulfill the tasks now asked of them without computers that solve problems for them at superhuman speeds. If you decided to "humanize" industrial, financial, commercial, and government operations by employing human beings in place of computers, you wouldn't have enough hands, let alone enough brains.

And if society grows still more complex over the next century, as, barring catastrophe, it surely will, then we will have to become still more computerized and automated than we are now.

Doesn't that mean there's a chance of abuse? Of course. There's a chance of abuse in everything. Think what could be done (and sometimes is) by a general who decides to abuse his military power.

Still, abuses don't take place more often than they do because all of society is the more stable the less abuse there is. It is to everyone's interest, therefore, to minimize abuse. Computers can supply the necessary techniques, perhaps, to monitor abuses more efficiently than ever before in history — even abuses of computers.

But if computers guard computers, who guards the guardians? That's an old question, and there *is* an answer. Each guardian keeps an electronic eye on all the others. In the American government system, we call that separation of powers.

What good will further computerization do us? One daydream involves a world without money.

All through history, financial transactions have been increasingly etherealized. Metal coins of universal use have replaced particular real objects. Paper bills, arbitrarily marked, substituted for coins. Checks, homemade bills of any denomination, substituted for government bills. Credit cards, a month's checks at a time, substituted for individual checks. The next step? Automatic electronic computation of your net worth, kept up to date with every transaction?

Suppose everyone possessed an appropriate device keyed

to her or his thumbprint, or to the chemical structure of her or his perspiration, or to something more subtle still. Suppose that at some appropriate manipulation, that device could tell you the exact state of your liquid assets — the amount you could use for transactions.

Any transaction you were involved with — earning money, investing money, spending money, from the purchase of a newspaper to the sale of General Motors — would become legal only when the devices of every party to the transaction were placed into a computer outlet, which would then transfer the necessary sums (in electron pulses) from one card to another. What we would have would be an endless procession of instantaneously written checks for any amount under the asset level.

The government could automatically take its share of money out of every transaction, basing its cut on the size of the deal and on the size of the asset level of the individual receiving the money. Other complexities would be taken care of and adjustments could be made, one way or the other, by the end of the year.

In such a cashless society, the concept of money and wealth would become less important because it would be less visible. This would be especially true if next century's society could find some logical way of lessening, at least somewhat, the unevenness of distribution of assets. Moreover, in such a cashless society, paying taxes would be less painful because the transactions would never be visible.

Abuses? They might actually decrease as dishonest dealing and tax evasion became more difficult. Part of the inhumanity of the computer is that once it is *competently programmed* and working smoothly — it is completely honest.

7 THE NEW CAVES

DURING the ice ages, human beings exposed to the colder temperatures of the time often made their homes in caves. There they found greater comfort and security than in the open.

We still live in caves called houses, again for comfort and security. Virtually no one would willingly sleep on the ground under the stars. It is possible that we may someday seek to add further to our comfort and security by building our houses underground — in new, man-made caves.

It may not seem, at first thought, a palatable suggestion. We have so many evil associations with the underground. In our myths and legends, the underground is the realm of evil spirits and of the dead, and is often the location of an afterlife of torment. (This may be because dead bodies are buried underground and because volcanic eruptions make the underground appear to be a hellish place of fire and noxious gases.)

Yet there are advantages to underground life too, and something to be said for imagining whole cities, even humankind generally, moving downward, of having the outermost mile of the Earth's crust honeycombed with passages and structures, like a gigantic anthill.

First, weather would no longer be important, since it is primarily a phenomenon of the atmosphere. Rain, snow, sleet, fog would not trouble the underground world. Even temperature variations are limited to the open surface and would not exist underground. Whether day or night, summer or winter, temperatures in the underground world would remain equable and nearly constant. The vast amounts of energy now expended in warming our surface surroundings when they are too cold, and cooling them when they are too warm, could be saved. The damage done by weather to man-made structures and to human beings would be gone. Transportation over local distances would be simplified. (Earthquakes would remain a danger, of course.)

Second, local time would no longer be important. On the surface, the tyranny of day and night cannot be avoided, and when it is morning in one place, it is noon in another, evening in still another, and midnight in yet another. The rhythm of human life therefore varies from place to place. Underground, where there is no externally produced day, but only perpetual darkness, artificial lighting that produces the day could be adjusted to suit humanity's convenience.

The whole world could be on eight-hour shifts, starting and ending on the same stroke everywhere, at least as far as business and community endeavors were concerned. This could be important in a freely mobile world. Air transportation over long distances would no longer have to entail jet lag. Individuals landing on another coast or another continent would find the society they reach geared to the same time of day as at home.

Third, the ecological structure would be stabilized. To a certain extent, humankind encumbers the earth. It is not only its enormous numbers that take up room; more so, it is all the structures it builds to house itself and its machines, to make possible its transportation, communication, and recreation. All these things distort the wild, depriving many species of plants and animals of their natural habitat — and sometimes, involuntarily, favoring a few, such as rats and roaches.

If the works of humankind were removed below ground — and, mind you, below the level of the natural world of the burrowing animals — humans would still occupy the surface with their farms, their forestry, their observation towers, their air terminals, and so on, but the extent of that occupation would be enormously decreased. Indeed, as one imagines the underground world becoming increasingly elaborate, one can visualize much of the food supply eventually deriving from hydroponic growth in artificially illuminated areas underground. The Earth's surface might be increasingly turned over to park and to wilderness, maintained at ecological stability.

Fourth, nature would be closer. It might seem that to withdraw underground is to withdraw from the natural world, but would that be so? Would the withdrawal be more complete than it is now, when so many people work in city buildings that are often windowless and artificially conditioned? Even where there are windows, what is the prospect one views (if one bothers to) but sun, sky, and buildings to the horizon — plus some limited greenery?

And to get away from the city now, to reach the real countryside, one must travel horizontally for miles, first across city pavements and then across suburban sprawls.

In an underworld culture, the countryside would be right there, a few hundred yards above the upper level of the cities — wherever you are within them. The surface would have to be protected from too frequent or too intense or too careless visiting, but however carefully restricted the upward trips might be, the chances are that the dwellers of the new caves would see more greenery, under ecologically healthier conditions, than dwellers of surface cities do today.

However odd and repulsive underground living may seem at first thought, there are things to be said for it — and I haven't even said them all.

8 THE LENT TO COME

NOTHING is ever used with complete efficiency. That's one of the consequences of what is called the second law of thermodynamics. In any engine some of the energy of the fuel is used up in overcoming friction, and a lot of it is just wasted as heat. We're extremely lucky if as much as 40 percent of it is actually used in running the engine.

In the same way, plants use only a small percentage of the energy of sunlight in forming tissue out of carbon dioxide, water, and minerals. (Fortunately, sunlight is plentiful and everlasting, and even at such low efficiency, plant life covers the world.)

An animal that eats plants does better but uses only about 10 percent of the energy in its food to form and maintain its own tissues. If a certain kind of animal maintains itself by eating a certain kind of plant, the plants must grow at ten times the rate the animals grow. There must at all times be ten times as much mass of the plant life as of the animal life.

If, for any reason, animals multiply past that one-to-ten level, some are sure to starve. If all of them eat wildly, the supply of plant life dwindles, and more of the animals are sure to starve. In the end, the ratio will have to be restored, at a lower level of total mass for each.

If an animal lives by eating another animal, again the eater can only have one tenth the mass of the eaten. Therefore, for every 100 pounds of plant life, you have 10 pounds of animal life that lives on the plants, and 1 pound of a second kind of animal life that lives on the first kind.

CHANGE!

You can have a long "food chain" of animals eating animals eating animals before you finally get down to plants. In such cases, the farther you move away from the plant end of the chain, the smaller the total mass of that kind of animal that can exist.

It is for that reason that plant-eating animals tend to be larger than meat-eating animals. The plant-eaters can afford it. The largest land-dwelling animals are all vegetarian.

In the sea, the largest animal that lives solely on other large animals is the sperm whale, a mainstay of whose diet is the giant squid. Larger than the sperm whale, however, are some of the great baleen whales, including the blue whale which can have a length of up to 100 feet and a mass of up to 150 tons. The blue whale is the largest animal that ever lived — ever! — and it lives on tiny plant-eating shrimp called krill, present in great quantities in polar waters.

One way an animal could transcend the limits of the mass it can attain is to cut out some of the steps in the food chain. Instead of eating large animals that live on other animals, it can eat small animals that live on plants. Better yet, it can live on plants directly.

Most animals have little choice in what they can eat (the koala bear can eat only eucalyptus leaves — no eucalyptus leaves, no koala bear) but the human being is fortunate in being omnivorous and able to live on almost anything.

Most human beings, given the choice, seem to enjoy eating meat — but even if their diet consists of plant-eating animals, that is wasteful. Remember that 100 pounds of plants support but 10 pounds of the animals we eat, and that, in turn, supports only 1 pound of us. If we were to eat the plants directly instead of feeding them to the animals first, those plants would support 10 pounds of us. In short, if human beings were all vegetarians, ten times as many of them could exist as could if they all ate nothing but meat.

That is why, when agriculture was first developed about

ten thousand years ago, and grains became the mainstay of the diet, the human population increased dramatically wherever farming came into being. That is why, today, nations that are very densely populated and insufficiently well-off to be able to import food must live on a largely vegetarian diet.

Some animals that form part of the human diet live on plant life that human beings can't eat, such as grass. This could nevertheless still be wasteful. Why should so much land be devoted to pasture when some of it, at least, could be used to grow grain?

The whole world faces this question now. The human population is increasing by 80 million individuals each year. By 2010, there may be as many as 8 billion people on Earth instead of the present 4 billion. How will we feed them all?

In the effort and the enormous drive to produce more food, we may no longer be able to afford the luxury of beef, lamb, pork, and chicken. Meat may become a rare and even, perhaps, a forbidden luxury, and all of humankind may find itself forced into a long Lent of vegetarianism, until such time as population is brought under control and human beings learn not to outbreed the food supply.

9 WHERE HAVE ALL THE MONSTERS GONE?

IN ANCIENT TIMES, people had a great deal of fun imagining the strange plants and animals that might exist in the mysteriously distant unknown regions of the world. There were unicorns and sphinxes and dragons and giant birds and man-eating trees and so on and so on.

But we've filled the world now, and it seems to hold no

mysteries anymore. To be sure, we are discovering new species of plants and animals every day, but they are unspectacular — new insects, new mites, new worms, new lichens. Are there any large animals that remain to be found?

As recently as 1900, in the secluded rain forests of the Congo, zoologists first became aware of the okapi, an antelope about five feet high at the shoulders. It proved to be a member of the giraffe family, the only one alive without the characteristic long neck.

A second find came still later and was even more startling. On the small Indonesian island of Komodo, a lizard was discovered in 1912 that was not just a lizard — it was the largest lizard known to exist. It can grow to a length of 10 feet and a weight of about 300 pounds. It can run quickly and can dispose of any wild pig it may meet. In fact, cases where these lizards have killed human beings have been reported. They are called Komodo dragons, and anyone meeting the beast suddenly might be forgiven if she thought for one moment that she had somehow slipped back into the age of dinosaurs.

These finds were made in very out-of-the-way places, however, and there is a strong possibility that there are no more spectacular discoveries to be made anywhere any longer, at least on land. Certainly no large, hitherto unknown, land animals have been discovered in seven decades.

What about the sea, though? The sea takes up more area than the land does, and its waters are opaque. Who can tell what may be hidden in the vast depths of the ocean?

One ancient myth, for instance, has come to life. In medieval times, there were frequent accounts of a monstrous sea creature called the kraken which could, with its long tentacles, envelop an entire ship and carry it down to destruction. The accounts were somewhat exaggerated, but they were not really false. In 1857, scientists were finally convinced of the existence of the giant squid, which could attain a length of as much as 65 feet and engage in mighty combats with whales.

A still more startling discovery was made in 1938 when a fish called the coelacanth was netted near the southern shores of Africa. It is a primitive, moderately large fish that lives in the deeper regions of the sea. It has stubby limbs fringed with fins, and probably hobbles about on them on the sea floor in search of food. The coelacanth is related to those fish which, 350 million years ago, emerged on dry land and became ancestors of all land vertebrates, including us.

The amazing thing about the netting of the coelacanth was that till then zoologists had been convinced the fish had been extinct for 60 million years! Finding a living dinosaur would not have been more surprising.

Might not the ocean hold more surprises: unknown giant creatures that might rival the whales and sharks in spectacular size?

Oddly enough, the most persistent reports of such a monster do not involve the sea, but a fresh-water lake that is not particularly large. It is Loch Ness in Scotland, a lake about 22½ miles long, 1 mile wide, and 750 feet deep in places. There have been numerous reports of sightings of some large animal, popularly known as the Loch Ness monster, that lives in its depths.

It is hard to take the reports seriously. So small a quantity of water, one would suppose, could scarcely support a population of large animals and hide them effectively.

Still, it is a dramatic possibility. The people who live near Loch Ness, aware of the tourist value of the monster, insist on its existence, and so do romantics everywhere.

In recent years, technology has joined the search. The depths of the lake have been studied by underwater photography and by sonar. Nothing suspicious has yet been detected, but believers have not lost their faith.

A new and particularly ingenious plan is being proposed

for the near future. Two dolphins are being trained to seek out large animals underwater and to use cameras strapped to them. The plan is to turn them loose in Loch Ness and have them stalk and snap the shy and fugitive monster.

So far, no such scheme has been put into practice. My guess (spoilsport that I am) is that when, and if, one is, it will find nothing.

10 THE TINY MAP

MEDICINE has more or less conquered infectious disease. The disorders that make us miserable and threaten us now, however, are not so simple. Today's dangers are born of failures in the fabric of our own chemistry.

No matter how we try, we can't seem to find out what exactly goes wrong to produce such disorders as arthritis, atherosclerosis, or cancer. We can't discover what precise chemical shifting of atom groupings within the cells starts the trouble, or makes it worse, or makes it better.

But then consider how complicated the human body is. The adult human being is made up of some 50,000,000,-000,000 (that's *fifty trillion*) microscopic cells of dozens of types, each containing 50 trillion atoms or more.

Each cell can, and often does, affect the others, and within each cell the constituent atoms are grouped into many thousands of different kinds of molecules, all engaged in chemical reactions that compete with each other for particular atoms and groups of atoms. Each particular reaction blocks or stimulates other reactions in a subtle way, and the whole forms an incredibly complex three-dimensional lacework that shifts constantly and rapidly, like the largest and fastest kaleidoscope you can imagine.

How can you plunge into this microscopic maelstrom and somehow pluck out that particular chemical reaction which, in its flash-by, goes just sufficiently awry to set up a hundred ill-working ripples that, in combination, initiate a cancer? Or how can one adjust the reactions in such a way as to end the cancer — except to try everything one can think of and hang on to whatever seems to do good for some reason we don't understand.

Perhaps we can cut away the nonessentials. Each chemical reaction within the cell is likely to be controlled by a particular molecule called an enzyme. If we knew all the enzymes present, and their relative proportions, we might be able to work out the reaction network. The enzymes, however, are very delicate molecules, hard to handle, and they're all mixed up in the internal chemical soup of the cell.

But then, each enzyme is formed by a gene, and each gene is a small section of a chromosome that exists within a small enclosed portion of the cell called a nucleus. The genes are made up of molecules of deoxyribonucleic acid or, abbreviated, DNA. If we knew all about DNA molecules, we might know all about the enzymes and, therefore, all about the busy chemical factory we call a cell.

As it happens, scientists are learning more about DNA all the time. There are now special techniques to cut the DNA molecules in chosen places and then recombine them. (This is what we call recombinant DNA.)

A molecule of DNA, once cut, can be recombined in some new fashion. A molecule can be modified; a part cut out; a portion of an alien gene inserted.

In this way, we might develop organisms with new prop-

erties. Bacteria have been developed that are capable of forming insulin, which they don't need but human diabetics do.

Even more important is the possibility that by the use of
recombinant-DNA techniques, we might learn to identify
genes along a chromosome and find out the exact structure
of each. We would then end with a "gene-map" of cells —
first of bacterial cells, perhaps, then of cells of higher organisms, finally of human beings.

By delicately modifying each gene and observing the
changes introduced into the cell in which it is located, we
might discover the enzyme associated with each gene and
what exactly that enzyme does.

Beginning, then, with the tiny map of the cell's genes,
taking into account each enzyme, and the nature and quantity of the simple chemical building-blocks in the cell, we
may then be able to begin to work out the network of reactions that takes place in the cell.

Once we have laid out a basic framework, we might be
able to program a sufficiently complex computer to mimic
those reactions. Working with the computer, we could then
create a model of the ever-shifting atoms and study, in detail, the equilibrium that exists.

Working with the computer rather than with the cell, we
could introduce changes in particular genes, which will bring
about modifications in particular enzymes, which will bring
about delicate shifts in the chemical equilibrium within the
cell. Adjusting matters this way and that, we could finally
shift the equilibrium to the point where abnormalities build
up that we will recognize as characteristic of this disease or
that. Given this knowledge, we can check back to the dis-
ordered cells to see if the computerized guess is correct.

If, in this way, we learn the particular change involved
in a particular disorder, we would surely have a better chance
of working out a scheme of attack that would prevent its
onset, or reverse it once it has started.

Might not cancer, atherosclerosis, arthritis, diabetes, and all the rest then go the way of smallpox and poliomyelitis?

---

11 THE OTHER YOU

WHAT IF our society uses new-found technologies of genetic engineering to interfere with the biological nature of human beings? Might that not be disastrous?

What about cloning, for instance?

Cloning is a term originally used in connection with non-sexual reproduction of plants and very simple animals. Now it is coming into use in connection with higher animals, since biologists are finding ways of starting with an individual cell of a grown animal and inducing it to multiply into another grown animal.

Each cell in your body, you see, has a full complement of all the genes that control your inherited characteristics. It has everything of this sort that was in the original fertilized egg cell out of which you developed. The cells in your body now devote themselves to specialized activities and, in many cases, no longer grow and differentiate — but what if such a cell, from skin or liver, could be restored to the environment of the egg cell? Would it not begin to grow and differentiate once more and finally form a second individual with your genes? Another you, so to speak? It has been done in frogs, and recently in mice, and can undoubtedly be done in human beings.

But is cloning a safe thing to unleash on society? Might it not be used for destructive purposes? For instance, might not some ruling group decide to clone their submissive, downtrodden peasantry, thus producing endless hordes of semirobots who will slave to keep a few in luxury and may even serve as endless ranks of soldiers designed to conquer the rest of the world?

A dreadful thought, but an unnecessary fear. For one thing, there is no need to clone for the purpose. The ordinary method of reproduction produces all the human beings needed as rapidly as they are needed. Right now, the ordinary method is producing so many people as to put civilization in danger of imminent destruction. What more can cloning do?

Second, unskilled semirobots cannot be successfully pitted against the skilled users of machines on farms, in factories, or in armies. Any nation depending on downtrodden masses will find itself an easy mark for exploitation by a less populous but more skilled and versatile society. This has happened often enough in the past.

But even if we forget about slave hordes, what about the cloning of a relatively few individuals? There are rich people who could afford the expense, or politicians who could have the influence for it, or the gifted who could undergo it by popular demand. There could then be two of a particular banker or governor or scientist — or three — or a thousand.

Might this not create a kind of privileged caste, who would reproduce themselves in greater and greater numbers, gradually taking over the world?

Before we grow concerned about this, we must ask whether there will really be any great demand for cloning. Would *you* want to be cloned? The new individual formed from your cell would have your genes and therefore your appearance and, possibly, talents, but *he would not be you.* The clone would be, at best, merely your identical twin. Identical twins share the same genetic pattern, but each has its own individuality and they are separate persons.

Cloning is *not* a pathway to immortality, then, because

your consciousness does *not* survive in your clone, any more than it would in your identical twin if you had one.

In fact, your clone would be far less than your identical twin. Genes alone do not shape and form a personality; the environment to which they are exposed helps. Identical twins grow up in identical surroundings, in the same family, and under each other's influence. A clone of yourself, perhaps thirty or forty years younger, would grow up in a different world altogether and be shaped by influences that would be sure to make him less and less like you as he grows older.

He may even earn your jealousy. After all, you are old and he is young. You may once have been poor and struggled to become well-to-do, but he will be well-to-do from the start. The mere fact that you won't be able to view it as a child, but as another competing and better-advantaged *you*, may accentuate the jealousy.

No! I imagine that, after some initial experiments, the demand for cloning will be virtually nonexistent.

But suppose it isn't a matter of your desires, but of society's demands.

I, for instance, have published nearly 250 books so far, but I am growing old. If there were a desperate world demand for me to write five hundred more books, I would have to be cloned. The other me, or group of mes could continue. Or could they?

The clones will not grow up my way. They won't be driven to write, as I was, out of a need to escape from the slums — unless you provide each with slums to escape from. Unlike me, they will all have a mark to shoot at — the original me. I could do as I pleased, but they will be doomed to imitate me and may very well refuse. How many of my clones will have to be supported and fed and kept out of trouble in order to find one who will be able to write like me and will want to?

It won't be worth society's trouble, I assure you.

12 YOUR SPARE PARTS

I HAVE JUST cast cold water on the notion of cloning as a way of producing a new individual like yourself.

But suppose we look into the matter a little further. Even if a cloned individual were to hold no interest for the world, cloning as a technique may nevertheless prove more versatile than I have suggested.

Consider how an individual develops. He or she starts from a fertilized ovum, which divides into two cells, each with the same genetic equipment as the original cell (the same genes, lined up in the same chromosomes). These divide into four cells, which divide into eight cells, and so on, all with the same genetic equipment. After a little over forty divisions of this sort, the fertilized ovum becomes a full-grown human being with some 50 trillion cells — all of them (barring accidental mistakes in the division process) with the same genetic equipment as the original fertilized ovum.

Yet the human being has a large variety of organs in his body — skin, liver, kidneys, heart, lungs, stomach, and so on — each of which has a different structure, behavior, capacity, and chemistry. How is this possible when all have the same genetic equipment, and the genes determine the chemistry and behavior of a cell?

Apparently, as the fertilized cell develops, different parts of the multiplying cell undergo differentiation. Certain of the genes are blocked out, leaving certain others active. The different types of partial activities guide one group of cells into developing as a liver, another group as a heart, still another as a brain, and so on. Naturally, each different type of cell loses some capacity for all-round activity. None of the body cells can serve, under natural conditions, as a beginning for a complete individual (except for the sperm and ovum, which are not, strictly speaking, *body* cells). Some of them are so specialized that, as in the case of brain cells, they cannot divide at all.

We don't as yet know the details concerning which genes are blocked to make one organ or another, or how the blocking is done; but we may learn as our knowledge of genetic detail increases. In cloning a body cell, then, we may be able to block, deliberately, just those genes of the nucleus that will cause the cell to develop into a heart only, or a kidney only, or a lung only . . .

Suppose every individual, soon after birth, has some fifty or so bright, young cells removed from the body, in a very minor operation, and has those cells preserved through freezing or some other technique. Suppose also that women generally contribute to banks of ova. (More than 90 percent of the ova women produce go to waste anyway.)

Once there are signs that a heart is beginning to fail, or a kidney, or a thyroid, a nucleus from one of the body cells can be inserted into an ovum in place of that ovum's own nucleus. The genes of the new nucleus would be unblocked so that it could begin developing outside the body in an artificial womb. At a given point, a set of genes would be blocked in such a way as to cause the cell to develop into a full-grown heart or kidney or thyroid, with the rest of the body, in each case, vestigial.

The technique of transplantation has already been worked out, but the great difficulty lies in the body's rejection of a foreign tissue. With an organ grown from a cell of the body itself, with identical genetic equipment, there is no problem of this sort. While autoimmune response (an allergy of the body to some aspect of itself) is not unknown, it is not common either, and your own new heart can replace your own old heart with very little trouble.

The body — like an automobile — would have its natural lifetime extended by the availability of spare parts. And think of the psychological effect! Each person will know that his or her heart, kidneys, lungs, glands are not all there is, that available replacements await the need. What a vast load of fear and anxiety would be removed.

Remember too that the new organs could supply new cells for new cloning. Assuming that all this works (and, of course, we can't be sure that there may not be insuperable difficulties that develop) would it mean that human beings will live forever?

Probably not. For one thing, not all breakdowns can be easily replaced. For another, in humans, as in automobiles, the increasing rate of breakdown may eventually make maintenance prohibitively expensive.

Finally, there is the problem of the brain. Whatever else is replaced, the brain will eventually deteriorate to the point where further repair elsewhere is useless. Nor can the brain be replaced, for even if a new brain is grown out of a body cell, it will contain only the genetic equipment of the original, and not the memories and experiences. With a transplanted brain, you are no longer you; and if a new individual is needed, it would be best to start from scratch in the usual way.

But even so, the bank of spare parts would at least mean you could live, with considerable youthfulness, till the brain goes, surely for a century or more, and that's much longer than most of us have now.

13 VARIETY!

LIFE is competition. This shows itself most clearly and dramatically when one animal kills and eats another. Even active killing aside, however, there is competition, continuous and ruthless, with quarter neither asked nor given, and with death as the end.

Animals compete for food, for space, for mates. Plants too compete for sunlight and water.

The competition powers change, putting a premium on the strongest and wiliest predator, on the most sensitive and fleetest prey, on those who can gather most or make do with least, on those who can strike out most boldly, or hide away most effectively.

So many strategies are possible that competition has resulted in the formation of several million species of plants and animals now alive (and ten times as many now extinct), each of which exploits its natural niche in some distinctive way, and each of which depends for life and well-being on the existence of the rest. The whole structure of interrelationships of species with each other and with the inanimate environment is the subject of the science of ecology.

With the coming of man, however, competition became dangerously one-sided. Humankind, plus its tools and weapons, was too powerful, and those plants and animals that most directly competed with it, went down; those that most directly served it went up.

The planet is now loaded with the weight of 200 million tons of living human beings — probably a larger weight than has ever before existed in the form of a single species, with an even larger weight of those species of plants and animals that serve humanity as food, slaves, or pets.

Other species are in retreat. Even in prehistoric times, magnificent animals faded before the master species. It required nothing more than spears and arrows to wipe out the mammoth some twelve thousand years ago, or the moa some three hundred years ago.

Nowadays, the larger species are rapidly shrinking in numbers. It is not so much that they are being deliberately killed as that their habitats are being wiped out by the remorseless expansion of humankind's numbers.

But if humankind is the first species to so overbalance the system of competition as to threaten the extinction of life's variety, it is also the first species to be willing to grant quarter. Quite apart from the danger to ourselves inherent in a too-drastic disruption of the ecological structure, we are coming to appreciate the aesthetic value of life-variety. Living species are usually too beautiful and too interesting to destroy. We want them all to live so that Earth can be a jewel of life, glittering in the universe.

But how are we to preserve the endangered species? One answer lies in the zoos of the world, which have become more than showplaces of interesting organisms. They now represent places where animals may be preserved in their last stand; where the remnant can be kept from extinction.

Another answer, which may develop in the future, rests with cloning (see the previous two essays). Where few pairs of individuals are left, and where extinction is a yearly possibility, indefinite numbers of new individuals can be produced by placing the nuclei of body cells of the threatened species into ova of similar but more populous species.

Of course, all the individuals produced from the cloned body cells of a single organism will be of the same sex and will be genetically identical. This would not be as good for the species as the genetic variety of a large natural population — but surely it is better than extinction.

A particularly dramatic suggestion of this type once reached me by mail. (Alas, I don't have the letter and can't give the name of the suggester.)

It deals with mammoths, the hump-backed, long-haired variety of elephant that roamed Siberia in the last Ice Age. They are occasionally found frozen in the tundra, and the meat is still fresh enough for dogs to eat without ill effect.

What if mammoths are found (at least one of each sex) with some cells capable of being thawed out into a spark of life? If a nucleus of such a cell could be placed into the ovum of a modern elephant, then, two years later, that elephant might give birth to a baby mammoth. It might then be possible to reconstitute the species and allow it to breed naturally.

And someday, when Earth has become a parklike world with most of the human population on space colonies, those species that have been preserved will yet roam the planet in such variety as remains.

14 HUMANITY ALONE

IN THE PREVIOUS ESSAY, I talked about the possibility of using cloning to save endangered species. The question is, though, can *anything* save endangered species, as long as human numbers continue to increase?

Every year there are more human beings on Earth than there were the year before; every year human beings take up more and more room on Earth and leave less and less room for other forms of life. Do you think that's nothing to worry about — that there's plenty of room, that we can always keep animals in zoos? Do you think that even if some of the larger animals become extinct — tigers and orang-utans, for instance — it makes no difference?

Let's think about it a moment, and begin with an interesting estimate — that about 20 trillion tons of living creatures of all kinds exist on Earth right now.

Most of this is plant life. It's got to be, because animals live on plants. (Even those species of animals that live on other animals are living on animals that, in turn, live on plants, or that live on still other animals that live on plants. No matter how long the chain of animals eating and being eaten is, it comes down to plants in the end.)

In general, when one type of life form does the eating, while another type is eaten, the total mass of the eaters cannot be more than about 10 percent that of the eaten. If the eaters proliferate too freely, then too much of the potential food is eaten and famine sets in. The eaters die off and the eaten recover.

On the whole, then, the total mass of animal life on Earth is roughly a tenth the total, or 2 trillion tons.

Some of that animal life consists of human beings. At the present moment, there are 4 billion human beings on Earth. If we allow for children and say that the average weight of a human being is 100 pounds, then the total mass of humanity is 200 million tons.

The total mass of animal life on Earth is ten thousand times as much as the total mass of human life on Earth. Since there are perhaps 2 million different animal species on Earth, the mass of human beings is about 200 times as great as it would be if every species contributed equally to the total mass.

And the mass of humanity is increasing. At the present

moment (1976), the Earth's population is mounting so rapidly that it will double in 35 years. If this actually happens, then the total mass of animal life on Earth will be only five thousand times as much as the total mass of human life in A.D. 2011.

If the doubling continues at that rate, in 470 years — in A.D. 2446 — the total mass of human life on Earth will be equal to the total mass of animal life on Earth. That would be the result of a little over thirteen doublings, starting at the present figure of the population.

That's a lot of doublings. Can we do it? After all, in the last ten thousand years, since the first stirrings of civilization, the human population on Earth has doubled only a little over seven times. Can we now double our population a little over thirteen times in less than five hundred years?

Even if we decide that humanity's scientific knowledge is increasing so rapidly that we can continue to feed the population and keep it alive no matter how often its numbers double — the next question is, do we *want* to double our population a little over thirteen times in less than five hundred years?

If the total mass of human beings becomes equal to the total mass of animal life in A.D. 2446, it will take all the plant life in the world to feed the vast numbers of human beings who will be alive. The total mass of humanity and the total mass of animals will be equal, and there will be no room, no room *at all*, for any animals other than human beings.

It's not just the tigers and orang-utans who will be gone. Cats and dogs will be gone. Sparrows and canaries will be gone. Frogs, fish, and grasshoppers will be gone. Every other animal species will be gone, and humanity will be left alone — in less than five hundred years at the present rate of increase — in a world given over, otherwise, to plants.

And not just to any plants. There can be no margin for inefficiency, and we couldn't allow plants that had inedible portions — bark, wood, cellulose. We would have to grow some kind of microorganism that was one hundred percent edible and live on that.

In A.D. 2446, then, there would be only human beings on Earth, together with the plant mush they would be living on, and no other life. Then that would be it — there would be no further room for multiplication.

Do you think we could make food synthetically, using solar energy, and no longer need plants? Maybe, but if we keep on multiplying at the present rate, by A.D. 3500 the mass of human beings would be equal to the mass of the Earth.

Do you think we could populate other worlds, or build space colonies, and no longer need the Earth? Maybe. But if we keep on multiplying at the present rate, by A.D. 6800, the mass of human beings would be equal to the mass of the known universe.

So we'll be forced to stop this insane increase in our numbers eventually, no matter what we do — and we had better make up our minds to do it right away. If we don't, then *we* are the endangered species.

NOTE: As I said in my note after Essay 4, "The Falling Birthrate," the birthrate *has* been dropping over the last decade — but not quickly enough as yet.

15 THE TINIEST SLAVES

BEFORE WRITTEN HISTORY began, human beings had already tamed all the animals and plants they were going to tame. Cats, dogs, pigs, goats, sheep, cattle, horses, donkeys, geese, ducks, hens, grains, fruit trees were all there, thousands of years ago, to be, in one way or another, our slaves: to be our companions, to do our work, to supply our stomachs with part or all of themselves.

More primitive animals have also been put to use. We plunder beehives for honey and certain caterpillar cocoons for silk. We can put yeast cells to work fermenting our fruit juices and our soaking grain. These things too date far back in history.

Is there nothing we can do now with the other species that share the Earth with us that our uncivilized or scarcely civilized forebears had not already done thousands of years ago?

Consider the bacteria. These are tiny living things made up of single cells far smaller than the cells in plants and animals. Four hundred trillion bacteria of average size (that's a hundred thousand bacteria for every human being on Earth) would weigh just a pound.

Bacteria exist all about us, and inside us as well, in vast hordes and in fifteen hundred different basic forms or species. A few of them cause disease, but the percentage is very low (far lower than the percentage of human beings who commit crimes). Most bacteria do us no harm, and many of them are essential to life.

Biochemically, the bacteria are amazingly versatile. No

naturally produced materials cannot be broken down by one type of bacteria or other. Decay bacteria restore everything to the biosphere to be used over again. If their work stopped, the world would become littered with undecayed scraps of indigestible matter, and these would accumulate till all life stopped.

Other bacteria can combine (or "fix") the nitrogen of the air with other elements to form substances that maintain the fertility of the soil. Without them, the soil and waters of the Earth would slowly grow sterile.

Bacteria can carry out chemical reactions that higher animals cannot, and we benefit by it. Bacteria in the cow's stomach digest the cellulose of grass and hay (the cows can't do it themselves). The products of digestion are absorbed by the cow and, eventually, come back to us as milk and meat. Bacteria in our own intestines form some of the vitamins we can't make for ourselves.

Well, then, can we improve the extent to which we depend on bacteria and force them to work for us more than they do? Can we make them into our tiniest slaves?

We may be able to. Like all other living things, bacteria contain nucleic acids that guide the formation of the bacterial enzymes, which, in turn, dictate the kinds of chemical things bacteria can do. Our molecular biologists are learning how to break nucleic acids in two and to recombine them in new ways. They are learning how to introduce nucleic acid pieces from one kind of cell into another. The result is that bacteria with new chemical abilities might conceivably be formed.

This is not entirely unprecedented in the human experience. On a larger scale, domestic animals have been bred to improve their natural yields of milk or eggs or wool; they have been bred larger and stronger, or smaller and cuter. Why not do the analogous thing with bacteria, and get some that specialize in some task we can put them to?

Diabetes, for instance, is a common disease, and diabetics need insulin if they are to live normal lives. Insulin comes from the pancreas of domestic animals — one pancreas per slaughtered animal, so that the supply is limited and cannot easily be increased.

Yet we have now "designed" a bacterium that can manufacture human insulin and the product is identical in every respect to the insulin our own bodies make. We can in time grow the stuff by the ton.

We might also design bacteria to manufacture other hormones; or to produce certain blood factors needed to clot blood, factors hemophiliacs lack; or to produce vaccines for use against those still smaller disease agents, the viruses.

We might also design bacteria that fix nitrogen more efficiently and rapidly, and in other ways improve the soil so that the agricultural yield can be doubled and redoubled. Or perhaps bacteria can be designed to make food out of sunlight as green plants do.

The scavenging activities of bacteria, already so versatile, might be improved. Suppose bacteria are developed that are exceedingly efficient at absorbing and metabolizing hydrocarbon molecules. They could be used to mop up oil spills — not only removing them from the environment, but converting them into protein, which, after a number of stages of eating and being eaten, will reach our own tables.

Bacteria might also be developed that can break down plastics that have been properly treated before being discarded and will not then act the part (as they do now) of undigestible, undecayable remnants. Bacteria can collect and concentrate traces of metals from wastes or from seawater.

In all these ways and others, we can, by properly designing these tiniest slaves of ours, use them to reshape the world itself and build it closer to our hearts' desire — but there are dangers too, as we shall see in the next essay.

16 OUT OF ADJUSTMENT

SUCCESSFUL PARASITES live amicably with their hosts. The most successful parasites do their hosts no perceptible harm and make their presence as undetectable as possible. Any parasite that harms its host is harming its own world and its own support. When the parasite goes to the extreme of killing its host, it dies itself. Those parasites best survive, then, that remain in careful adjustment to their hosts.

Each human being, for instance, is loaded with bacteria and other infectious agents which, by and large, live on us or in us without essential harm, charging us no more than we can afford, and held in check (just in case) by the body's natural defenses. Sometimes the situation goes out of adjustment and we develop infections, fevers, illnesses. Even then we usually recover, and affairs move back into adjustment.

The adjustment is not, of course, a matter of deliberate choice, but is the product of the cell chemistry that permits the parasite to attack and the host to defend in a kind of balance of power. The balance is attained by the processes of natural selection, since those parasites too efficient on the attack and those hosts too deficient on the defense tend to die off more quickly than the rest.

But cell chemistry can change from generation to generation through spontaneous changes or mutations in the genes that control that chemistry. Microscopic parasites reproduce so frequently and in such vast numbers that the total numbers of mutations are large, and some are sure to be threatening. A particular parasite may just happen to undergo changes that increase its capacity to attack, or to shift from host to host. A new strain of an old tame disease may suddenly become wildly virulent or very contagious, or both. Eventually, those hosts least capable of defense and those parasites least capable of restraint, die off, and a new adjustment is reached.

In human history, the worst such case came in the fourteenth century. Somewhere in Central Asia, about 1330, a new strain of plague bacillus appeared, one to which human beings were particularly defenseless. It spread and, by 1347, had reached western Europe. The disease was called the Black Death, and altogether it may have raged for a quarter of a century. In that time it managed to kill one third of the human species — 25 million deaths in Europe alone. It was the worst disaster ever to strike humanity.

There were other serious, worldwide epidemics, both before and after the Black Death. The most recent was the influenza that swept the world in 1918, just as World War I was ending. In one year it killed 30 million people. However, that represented only one-sixtieth of the world population of that time.

We tend to think of such dreadful visitations as things of the past. What with our knowledge of hygiene and our use of antitoxins and antibiotics, the natural defenses of the human body have been reinforced by the powerful defenses of modern medical technology. At the mere hint that 1976 might see an attack of influenza similar to the 1918 epidemic, plans at once began to inoculate every single American against it.

But if science can add to the defenses of the human being, it can also add, inadvertently, to the powers of the parasite.

In the preceding essay, I mentioned the ability of biologists to alter the genes of microorganisms. What they are doing is producing artificial mutations. Naturally, they hope to help themselves understand the machinery of life and to (possibly) produce new strains of organisms with properties that might be helpful to human beings.

It is not always possible, however, to predict the results of remodeling a gene. What if you produce the kind of microorganism that has an increased capacity to produce disease, or an increased ability to circumvent the natural defenses of the body, or an immunity to antibiotics? What if some ordinary, harmless microorganism that infests everyone is changed into a new and dangerous strain? What if some virus, after remodeling, turns out to possess the ability to alter the chemistry of the cells it invades in such a way as to start a cancer?

The chances of anything like this happening are very small. If it should happen, precautions are taken so that the chance of any such strain escaping from the laboratory are very small.

The danger, however, is so horrible — a new and more harmful Black Death, a rapidly spreading plague of cancer — that no matter how small the chances are of anything happening, even those small chances seem too large to be tolerable. There is a natural desire to make those chances zero by putting an end to any kind of research that involves tampering with genes in this fashion.

Yet most scientists are reluctant to bang shut this particular door of research. Give up the knowledge it might bring? Give up the useful applications that might result?

Actually, there is an answer to the dilemma and to others like it, which I will describe a little later.

17 BOY OR GIRL?

EACH WOMAN in her lifetime produces a number of egg cells that are all genetically identical. Every one has twenty-three chromosomes, including a rather long one called the X-chromosome.

Each man, however, produces two kinds of sperm cells. One contains twenty-three chromosomes, including the X-chromosome. The other contains a stubby Y-chromosome in place of the X. Each man produces equal quantities of each variety, and the odds are about equal as to whether an X-chromosome sperm or a Y-chromosome sperm will fertilize an egg cell.

If an X-chromosome sperm does the fertilizing, the fertilized egg that results is XX and develops into a girl. If a Y-chromosome sperm does the fertilizing, the fertilized egg is XY and develops into a boy.

As a result of the even breaks of the game, boys and girls are born in roughly equal numbers. About half the human beings on Earth are, therefore, men, and half are women.

As the techniques of human genetic engineering are developed, one of the early victories is likely to be the discovery of ways to tip the odds. It may become possible to allow a couple to choose whether they will have a boy or a girl, instead of settling for whatever the fates allot.

What will then happen?

For a variety of social and historical reasons, most couples seem to want baby boys rather than baby girls. The general feeling might be, then, that if every couple could choose the sex of their children freely, boys will come to predominate in numbers; perhaps overwhelmingly so.

Yet would this be a permanent situation?

Even if, in the first generation after free choice becomes possible, boys become common and girls scarce, would this not be an unstable situation? If girls were relatively few in number, would they not become relatively valuable for a number of reasons? Couples would then be more likely to want girls, and the pendulum would swing back. If the pendulum overshot the mark, it would be the stock of boys that would rise. The generations might zigzag some distance this side or that side of the fifty-fifty mark, but the ratio is not likely to stay permanently on either side.

But what if the question of the sex of children is *not* a matter of voluntary decision? Suppose the government takes a hand.

We might wonder why it should. Is there any reason why those planning the future of the species should decide to unbalance the ratio of the sexes? Why should they suddenly announce, say, that the quota of females was filled and only boy babies would be licensed for the remainder of the year?

Well, the birthrate depends on the number of women of child-bearing age in the population. The number of men, within broad limits, does not matter. One man could keep a woman pregnant all the time; two men (or any number of men) could not make her pregnant more often.

Therefore, if there were some way of controlling the sex of the unborn child, governments might deliberately enforce a deficit of females as a way of lowering the birthrate. We might even imagine a world government, at every census period, carefully weighing the statistics of population growth rates and setting the male-female ratios for the various regions of the world in such a way as to make those growth rates move up or down as desired.

On the other hand, suppose the principles of feminism are successful. Suppose it turns out that women are, for one reason or another, more able to withstand the rigors of space, or more able to adjust to a computerized world, or to a decentralized low-population world, than men are.

It might pay, then, to adjust the sex ratio so that there are ten women for every man. There would still be enough men to take care of the reproductive necessities, if social customs were altered to suit, or if men were used as sources of sperm for artificial insemination.

Yet is either extreme practical, whatever the uses? In a world in which men will have to share women or do without, or where women will have to share men or do without, might there not be too much squabbling to give people time for anything else? Harem politics (the worst kind) would predominate either way.

After all, the human species is the product of long ages of evolutionary history, and if its reproductive strategy involves equal numbers of the two sexes, I think it is because that works best. I suspect that any other arrangement will never be generally popular.

18 IT'S ALL IN THE MIND

PAIN SERVES a purpose. It tells you something is wrong and must be attended to.

But what if you *can't* attend to it for a while? Or what if you *have* attended to it and there's nothing more to do but wait.

The pain then keeps right on going even though it serves no further purpose, like a telephone that keeps on ringing even after you've picked up the receiver.

Naturally, there's a strong desire to find a way to end the

pain and, long ago, people found out how, more or less accidentally.

Opium, the dried juice from the unripe fruit of the opium poppy, was known in ancient Assyria nearly three thousand years ago. Chewing away at it reduced pain, misery, weariness and gave one a sense of well-being. Similarly, the people of South America chewed the leaves of the coca shrub long before the Europeans arrived.

In 1806 the chemical morphine was isolated from opium, and it turned out to be far more efficient as a pain reliever and calming agent than opium itself. Cocaine was isolated from coca leaves and used as a local anesthetic in 1884.

The trouble with these pain relievers is that they are addictive. The body gets accustomed to them, so that larger amounts must be used to produce the same effect. The body becomes *so* accustomed to the pain relievers that the time comes when it cannot easily do without them. Stopping their use then produces severe withdrawal symptoms. (Heroin, a modified form of morphine, is an even more efficient painkiller than morphine, and is even more dangerously addictive.)

These painkillers work by attaching themselves to certain areas (receptors) in the brain that receive those nerve impulses to which the brain ordinarily responds by producing the sensation of pain. The painkillers coat the areas and prevent the impulse from reaching the brain.

Sometimes people, under the stress of strong emotion, do not feel pain when ordinarily they would. Some natural chemical must block the pain receptors on rare occasions.

In 1975 such chemicals, named endorphins, were found and isolated from the brains of animals. The simplest endorphins have molecules built up of five smaller units called amino acids. Chains of amino acids make up the protein molecules of living tissues, but protein molecules are made up of chains of scores, hundreds, even thousands of amino acids. Very short chains are called peptides.

Even five amino acids offer the possibility of surprising variety, however. There are twenty different kinds of amino acids in protein molecules. If each of the five amino acids in a peptide can be any one of twenty, the total number of different peptides possible is 3,200,000. Each peptide has a somewhat different shape, and the endorphin shape just happens to fit the pain receptors.

The endorphins have a flaw as painkillers. The brain forms them as needed and can get rid of them efficiently by pulling the amino acids apart. When the endorphins are not the brain's idea, so to speak, but are supplied from outside, they are rapidly broken down in the brain, and the painkilling action does not last long. Morphine, on the other hand, is a molecule that is not easily broken down, and its effect is longer lasting.

Still there is the hope that if endorphins can somehow be made to work, they may not prove addictive, since the body deals with them naturally.

Indeed, we might wonder what else there might be in brain tissue. What other peptides might be present?

If millions of different peptides are possible with five amino acids, tens of millions are possible with six amino acids, and over a billion with seven.

We might find a few dozen peptides with different kinds of natural effects on the brain. We might even find ways of modifying those peptides slightly to make them more efficient.

It may even be, perhaps, that serious mental disorders such as those lumped under the heading of schizophrenia are the result of the overproduction or underproduction of some of these brain peptides. In that case, we might learn to diagnose and treat mental disorders in chemical terms and, in this way, empty the mental hospitals.

That may be a long way off, but it's a beautiful thought and worth striving for.

19 DIRECT CONTACT

THERE MUST BE millions of people who believe that telepathy is possible and that, in numerous cases, it has already been demonstrated. Yet, under present conditions, it does not seem at all likely that telepathy can exist. Consider —

In 1924, Hans Berger, an Austrian psychiatrist, placed electrodes against the human scalp and found that by using a very delicate galvanometer he could just detect a wavering electric potential. This seemed so far-out that he didn't dare publish his findings till 1929.

The technique has been refined since then. Changing electric potentials can be recorded as wavy lines on paper. This is called electroencephalography (from Greek words meaning electric brainwriting). It is abbreviated EEG.

The EEG usually shows certain rhythms. When you are asleep, there are large, slow delta rhythms. When you are awake, but keep your eyes closed, the rhythm speeds up into alpha rhythms. When you open your eyes, the rhythm speeds up further and grows smaller, as beta rhythms. Occasionally, small, slow theta rhythms are observed.

It is hard to tell what these rhythms mean, and why the electric potentials make themselves felt in the particular wavy variations they do. Every nerve cell in the brain transmits a tiny electrical impulse as it works, but there are something like 100 billion brain cells altogether. With each one contributing, all we can detect is an average.

Using an EEG is something like listening to human activity the world over. All we would hear would be a vague buzz. There would be rhythms. The buzz would be louder by day in any particular spot, softer by night. There would be subtler rhythms indicating evening hilarity, twice-daily rush hours, and so on.

From such an overall average, only large and serious disruptions can make themselves felt. EEG records are useful in diagnosing brain tumors or epilepsy. (In the same way, if we were listening to the human noise of the world, we might detect the existence of a war.)

What about individual thoughts in the brain, however? Could someone else detect them directly?

That would be like listening to the world's human activity and trying to detect something said by a single individual. The single comment would be drowned in the general noise — and so would the single thought in the brain.

For this reason, one must be skeptical concerning reports of telepathy. It is difficult to see how specific messages can be sent and received electromagnetically against the general background of brain noise, especially when the brain's overall electrical potential is only in the millionths of a volt. Perhaps some form of radiation we have never discovered and are not aware of is involved, but that is hard to accept too.

And yet — computers are becoming more and more versatile. Is it possible that the vague brain rhythms might be analyzed by computers far more delicately in the future than can be done now? Can the EEG record be split into many subsidiary rhythms, some of which are the result of thought processes, and can individual thoughts leave their mark there?

If so, it might be possible to amplify these particular rhythms, specifically, by some clever electronic device we have not yet invented. Such an amplified rhythm might produce a pulsing electromagnetic field that could be sensed over some reasonable distance. Someone else, with a similar electronic device, could receive a field that imposes itself on his brain, so that he becomes aware of a thought that not he himself, but someone else, has originated.

Telepathy might then become possible, through some appropriate device, though not by the unaided brain — just as we can now talk to others thousands of miles away, by telephone or radio, but not by the unaided voice.

Such telepathic ability would involve problems. How could such thought processes be made private so that you reach only the person you want to reach and not everyone around you? How could you activate someone's receiver and let him or her know you wanted to send a thought-message? And what advantage would it have over ordinary speech?

It seems to me that telepathy might be the answer to the language problem. Saying a word and thinking an image might be the ideal way to teach a language. In fact, in a pinch, thinking images might be sufficient in itself.

Of course, in telepathic communication, it would be a lot harder to self-censor impolite or impolitic thoughts that ordinarily exist inside us without our putting them into words. But then, perhaps human beings will get used to a new and greater candor. It may even do us good.

20 THOUGHT CONTROL

IN OUR technological civilization, we have been going through a process called etherealization, which means we have to apply less and less direct force to achieve a given result

Step 1. We use our hands, feet, and other parts of the body to accomplish a task, such as heaving a rock out of place.

Step 2. We use simple machines to direct or amplify the force, as when we use a lever to pry up a rock; or we use simple machines to reduce the obstructive effects of friction — as when we pull loads on wheeled vehicles instead of dragging them along the ground.

Step 3. We use an inanimate force to do the work, and confine our own efforts to controlling that force. We don't heave at a cable to lift an elevator; we simply shift a lever that closes an electric circuit that starts a motor that does the lifting.

Step 4. We find still easier ways to signal the inanimate force that does the work. We step on the gas and turn the ignition key instead of turning a crank. We push a switch or press a button instead of turning a lever. Why push or press anything, come to think of it? Why not just touch?

This fourth step is familiar to us in many cases. In large cities, we may signal elevators dozens of times a day by merely touching an object lightly. It illuminates to indicate that the touch was received, and the elevator comes at our bidding. Inside, we touch an appropriate place, the light goes on, and the elevator stops at the proper floor.

Why not use touch for the control of machinery generally? We can imagine a touch-console with machinery reacting according to the particular place or combination of places touched. The most complicated machinery could then respond to the virtuoso control of a master pianist, so to speak.

The question is, now, can etherealization progress still further? Is there anything simpler than the touch of a finger?

Why not the sound of a voice? It is much easier to say something at ordinary speaking level than to move the hand, or even the finger. It is not hard to imagine computerized machinery responding to the sound of a voice and reacting properly to the characteristics of different sounds: Stop! Go! Faster! Slower! Right! Left! Up! Down!

In fact, the voice lends itself to greater precision than the finger does. Every finger touch can control a machine equally well, unless we imagine the controls responding to a finger-print. Voiceprints are more easily responded to. After all, we ourselves have to study fingerprints very closely to be able to identify one, but any of us can recognize a voice, from a single word sometimes, even when that word is distorted by telephonic reproduction.

But is all this a sign of the growing laziness of human beings? And is it morally wrong to seek a reduction of effort?

Not really. By cutting down on human effort and shifting it to the inanimate world, we can do more — better, more quickly, and more precisely.

Besides, it is nothing new. All through history a few lucky individuals have been able to manipulate machines by kicks, pushes, or even just the sound of the voice. The machines happened to be two-legged animated beings called "slaves," "servants," or "employees." To say, "Jane clear the table," and have it done is etherealization for the few at the expense of the many. Why should it be immoral to have it done for the benefit of all?

Can we go even further than the voice in etherealization?

In the previous essay, I talked about the possibility of telepathy — the amplification of the tiny electromagnetic fields of the brain cells in such a way that the slight varia-tions presumably involved in thought could be impressed on other brain cells to establish communication. (We don't know how to do this right now, but it is conceivable, at any rate.)

Might we not use the same system to impress a thought on a computer designed to receive it? What is more ethereal than thought control? Think and the machine responds! What's more, each person's thought-patterns may well be as distinct as his or her voice, so that a machine might respond only to the thoughts of authorized personnel.

Of course, it would be troublesome to have idle or casual thoughts operate a machine inappropriately (like accidentally brushing a contact). We could suppose that someone would have to think some password first and *then* direct the machine.

By the time machines respond to thought, they may be capable of thought themselves and object to some orders, giving their reasons. If machines are not affected by human emotions and follies, it might pay us to listen to them. They may well turn out to be right.

21 THE FINE PRINT

THE FIRST important communications satellite went up in 1965. It, and others that followed, are owned and operated by International Telecommunications Satellite Organization usually referred to as Intelsat.

There are now communications satellites in stationary orbits (ones that allow them to circle the Earth in twenty-four hours, keeping step with some particular spot on Earth) over each of the three major oceans. They not only relay television programs from continent to continent, but they handle telephone traffic too. By 1975 perhaps one-third of the transatlantic telephone calls were going by satellite.

But transatlantic telephone traffic is going up at a rate of 20 percent per year and shows every sign of continuing to do so for a while. Intelsat will have to put up satellite after satellite and make the ground stations more numerous and more advanced to handle the traffic; or they will have to make it possible for individual satellites to handle more messages.

There is only so much room for communication in a given range of radio waves. You can send one message on a narrow band of wavelengths, a second on another narrow band, and so on, until you've occupied the entire range.

If you move into ranges with smaller and smaller wavelengths, there is more and more room to place different bands of wavelengths side by side. It is precisely as though you were sending printed messages and using smaller and smaller print — therefore getting more and more messages onto a given piece of paper.

Eventually, we should get into the tiny wavelengths of visible light and ultraviolet radiation, and we would then use laser beams to carry messages. Since a lightwave is something like a millionth as long as a typical radio wave, you can squeeze a million messages into a beam of light side by side across the same range that would carry one message by radio waves. Lightwaves are the equivalent of microprint.

In space, light offers no difficulties, but on Earth itself there are technical problems. Radio waves go through or around material obstructions, through mist and cloud and fog and dust; light is stopped by all of them. Work, however, is being done on optical fibers — thin threads of very transparent glass that carry light, like a metal wire carrying electricity. Perhaps, eventually, Earth may be "wired" with such optical fibers, and then —

A world in which every light beam can carry many millions of messages would be a revolutionary one indeed. Every person might have his or her own wavelength, equivalent to the personal telephone numbers we now have. Any person could reach any other person anywhere on Earth since, to satellites, all parts of the Earth are equally in view. If you had a projector that could be made to emit your own personal wavelength in case of need, you would never be lost, for the wavelength, inevitably picked up by search circuits, would not only tell rescuers where you were but who you were.

Closed-circuit television would be readily available for almost any kind of use because channels would exist in indefinite numbers. You could use picture-phones (if you were willing to be seen) or you could use your television screen for nonsocial purposes.

You could dial a newspaper, survey its pages, and have a facsimile printed of any page or section you wanted to study more closely. You could receive your mail, or sales information from supermarkets, by television. For that matter you could send mail by television or dial an order to the supermarket.

The libraries of the world might be so coded that any book, magazine, pamphlet, or document could, at the punch of the proper code, appear on the television screen for short-term reference — be produced, in part or in whole, in microform for long-term reference or in ordinary form for reading pleasure.

(I am enough of a writer, however, to insist, in this particular vision, that authors and publishers be paid a fee for each such use of their works.)

People could supervise machines and factories, or program computers that could run the world's work, by television. People could hold conferences by closed-circuit television, with each participant actually present in his or her own home.

Information, entertainment, instructions by lightwave; so much by lightwave that human beings can spread out at last. There might be no more purpose in huddling together as the only way of reaching one another. The whole world would be available from any point. It could become a large village composed of small villages and be utterly different from anything we have seen yet — thanks to the fine, fine print in beams of modulated light.

22 THE PUSH-BUTTON LIBRARY

IN THE next few essays, I will discuss a possible teaching-machine future. That is a far future, however. What will the nearer future be like in this respect? What is the first stage in reaching that far future?

As it happens, we already have a kind of teaching machine — the public library, which is the repository of the world's knowledge (or as much of it as a particular library can afford to have in a world in which almost anything from a battleship to a sweepstakes ticket is valued more than teaching) and the question is, how will it adjust to the future?

It will be more and more computerized, as obviously everything will be, assuming that our civilization suffers no catastrophe. More and more of the information will be on microfilm; more and more of the microfilm will be keyed to codes that can be entered into computer memories. More and more computers will then be used to come up with books, pamphlets, and articles that deal with the subject matter in question. There will first be titles and brief descriptions of contents, then a finer tuning to get down to particular items and to have the actual pages flipped and, perhaps, reproduced, if necessary.

(It should all be paid for, of course, for people have gathered and written and classified and computerized the material and they should be recompensed, but payment could well be out of public tax money on the grounds that the availability of knowledge is as much a human right — and, in the long run, a more valuable one — as police protection is.)

Now let's look at it the other way around. More and more homes are going to be equipped with computer terminals that can, for moderate sums, be keyed to information sources dealing with weather, news headlines, shopping services — or general information available in computerized libraries. For households that do not have terminals, there will be computer outlets in special retail establishments where, for a moderate rental (so much per minute) anyone with a few coins to spare can tap into the accumulated knowledge of the world. (There are already laundromats and photocopying establishments. Computer outlet rental shops will be much the same.)

How will this affect youngsters? Easy. Homework will, for the first time in history, become exciting — perhaps the big thrill of the day. It will be done on the computer outlets and the enterprising teacher should ask the students to look up something unusual that is related to the subject of the day and deliberately make the request a loose one.

If multiplication is on tap, who was the first one to use X to stand for the multiplication process, and why use a symbol anyway? If the steam engine is mentioned, who were the first to develop the steamboat and the steam locomotive and what happened to them? If the Magna Carta is being studied, what did King John's signature look like and why?

In addition to grading the homework in all the usual ways, there could be a special commendation for anything unusual that a student can bring in (with references) related to the subject, or even rather distantly related; something, perhaps, that the teacher didn't know or that would illuminate the day's study in an unusual manner.

The advantages?

First, the students will have fun. They will have a license, so to speak, to rummage through a push-button library.

Second, they will learn to use a computer in the most

flexible possible way and, more and more, life is going to be a matter of computers.

Third, they will learn about the resources of a library and the method and value of using it.

Fourth, they will be encouraged to be creative, for there is a reward for going off in odd directions — something that the proper use of the computer will make easy.

And fifth, they will learn the value not only of learning but of teaching, for if they are sufficiently innovative and clever in tracing down esoteric information, they will be able to inform the class, and the teacher too, of matters concerning which they might otherwise know nothing.

Many people have always felt that the public library, one of the great ornaments of the community, is far too likely to be the first to feel the pinch of shrinking funds and the last to feel the benefits of expanding funds.

Perhaps when the library is brought into the home and when children become avid to look things up and find novelty — moving on to things that interest them exceedingly even when it has nothing to do with schoolwork — we will all take a new look at a completely new kind of library, and that library will at last come into its own.

23 IT'S HOW YOU PLAY THE GAME!

Now THAT microcomputers are becoming more and more common, one hears doleful predictions that human minds will atrophy as all its functions are given over to machines. (I considered that very problem in "The Feeling of Power," a story I wrote on July 31, 1957, when computers were still the size of walls.)

We've faced the problem before, though. When writing was invented, five thousand years ago, there must have been many who thought that human memory would shrivel if all records were converted into marks on paper. Well, extraordinary feats of memory are indeed no longer necessary, but surely we've gained much more than we've lost.

The invention of the yardstick and the sundial limited the need to make clever estimates of distance and time. The invention of a symbol for zero, of Arabic numbers, of slide rules and cash registers, all, in one way or another, wiped out some of the requirements for mental ingenuity, but they have all given more than they have taken away. And they haven't taken away much. We have retained enough for our needs.

Computers, however, threaten to take away almost all the mental work with which we have been keeping our brains fit through the ages, leaving us nothing to do.

In the first place, computers take away the dull, routine, stultifying repetitious work that has been *ruining* our brains, not keeping them fit. In the second place, they leave us leisure, amusement, and creation.

For instance, computers can be programmed to play games — chess, for example. Though chess has never been completely analyzed, programs based on general principles make it possible for computers to play a passable game.

Once a person learns the rules of the moves, he or she can begin to play at once and, of course, be soundly trounced by the computer. Human players can, however, learn from their own mistakes and improve their game — more effectively against a computer than against another person.

A computer, after all, does not get tired, or impatient, or contemptuous, or busy with other things. It can be used at will and at the person's own pace. Eventually, the human player will learn to win, and the computer can then be reprogrammed more elaborately, to become a stronger adversary against whom the human player can continue to improve.

Will human activity become trivial, then? Will the great feats of the human mind vanish and will it be reduced to playing chess?

That's not the way to look at it. It is not that human beings are *playing* chess, it is that they are *learning* chess. Computer games are a device for teaching, and if a computer can teach you chess, it can teach you other things as well.

After all, human activities can all be interpreted in some ways as a game in which the adversaries are people, or nations, or economic groups. In the game of scientific research, humanity plays against the universe.

Such things can be treated in a branch of mathematics called game theory. Right now, game theory can scarcely handle very complicated aspects of such activities, but with more and more advanced computers, enormously complex "games" may entirely alter and *improve* the human ability to deal with the universe and with each other on the very deepest level.

But will this preoccupation with computers produce a society of isolates who will relate only to machines and never to each other? (I discussed this problem in my novel *The Naked Sun*, published in 1957.)

Why should it? In the first place, we don't need advanced machines to achieve isolation. People can and do lose themselves in books, in records, in television.

On the other hand, there can be a kind of rebound. Thousands of people were so utterly lost in the television program "Star Trek," for instance, that they began to seek out others like themselves to form fan clubs and hold conventions.

In short, what seems to be a force for isolation can actually become a pull toward human interaction. With everyone playing some sort of intense game, and looking for others equally interested, the result could be an intellectual ferment such as the world has never seen.

24 THE NEW TEACHERS

IN THE fifth essay of this book, "Adult Education," I discussed the fact that the percentage of older people in the world is increasing and that of younger people decreasing, and that this trend would continue if the birthrate should drop and medicine continue to extend the average life span.

In order to keep older people imaginative and creative and to prevent them from becoming an ever-growing drag on a shrinking pool of creative young, I recommended that our educational system be remodeled and that education be considered a lifelong activity.

But how can this be done? Where will all the teachers come from?

Who says, however, that all teachers must be human beings or even animate?

Suppose that over the next century communications satellites become numerous and more sophisticated than those we've placed in space so far. Suppose that in place of radio waves the more capacious laser beam of visible light becomes the chief communications medium.

Under these circumstances, there would be room for many millions of separate channels for voice and picture, and it is easy to imagine every human being on Earth having a particular television wavelength assigned to her or him.

Each person (child, adult, or elderly) can have his own private outlet to which could be attached, at certain desirable periods of time, his or her personal teaching machine. It would be a far more versatile and interactive teaching machine than anything we could put together now, for computer technology will also have advanced in the interval.

We can reasonably hope that the teaching machine will be sufficiently intricate and flexible to be capable of modifying its own program (that is, "learning") as a result of the student's input.

In other words, the student will ask questions, answer questions, make statements, offer opinions, and from all of this, the machine will be able to gauge the student well enough to adjust the speed and intensity of its course of instruction and, what's more, shift it in the direction of the student interest displayed.

We can't imagine a personal teaching machine to be very big, however. It might resemble a television set in size and appearance. Can so small an object contain enough information to teach the students as much as they want to know, in any direction intellectual curiosity may lead them? No, not if the teaching machine is self-contained — but need it be?

In any civilization with computer science so advanced as to make teaching machines possible, there will surely be thoroughly computerized central libraries. Such libraries may even be interconnected into a single planetary library.

All teaching machines would be plugged into this planetary library and each could then have at its disposal any book, periodical, document, recording, or video cassette encoded there. If the machine has it, the student would have it too, either placed directly on a viewing screen, or reproduced in print-on-paper for more leisurely study.

Of course, human teachers will not be totally eliminated. In some subjects, human interaction is essential — athletics, drama, public speaking, and so on. There is also value, and interest, in groups of students working in a particular field — getting together to discuss and speculate with each other and with human experts, sparking each other to new insights.

After this human interchange they may return, with some relief, to the endlessly knowledgeable, endlessly flexible, and, most of all, endlessly patient machines.

But who will teach the teaching machines?

Surely the students who learn will also teach. Students who learn freely in those fields and activities that interest them are bound to think, speculate, observe, experiment, and, now and then, come up with something of their own that may not have been previously known.

They would transmit that knowledge back to the machines, which will in turn record it (with due credit, presumably) in the planetary library — thus making it available to other teaching machines. All will be put back into the central hopper to serve as a new and higher starting point for those who come after. The teaching machines will thus make it possible for the human species to race forward to heights and in directions now impossible to foresee.

But I am describing only the mechanics of learning. What of the content? What subjects will people study in the age of the teaching machine? I'll speculate on that in the next essay.

25 WHATEVER YOU WISH

THE DIFFICULTY in deciding on what the professions of the future would be is that it all depends on the kind of future we choose to have. If we allow our civilization to be destroyed, the only profession of the future will be scrounging for survival, and few will succeed at it.

Suppose, though, that we keep our civilization alive and flourishing and, therefore, that technology continues to advance. It seems logical that the professions of such a future would include computer programming, lunar mining, fusion engineering, space construction, laser communications, neurophysiology, and so on.

I can't help but think, however, that the advance of computerization and automation is going to wipe out the subwork of humanity — the dull pushing and shoving and punching and clicking and filing and all the other simple and repetitive motions, both physical and mental, that can be done perfectly easily — and better — by machines no more complicated than those we can already build.

In short, the world could be so well run that only a relative handful of human "foremen" would be needed to engage in the various professions and supervisory work necessary to keep the world's population fed, housed, and cared for.

What about the majority of the human species in this automated future? What about those who don't have the ability or the desire to work at the professions of the future — or for whom there is no room in those professions? It may be that most people will have nothing to do of what we think of as work nowadays.

This could be a frightening thought. What will people do without work? Won't they sit around and be bored; or worse, become unstable or even vicious? The saying is that Satan finds mischief still for idle hands to do.

But we judge from the situation that has existed till now, a situation in which people are left to themselves to rot.

Consider that there have been times in history when an aristocracy lived in idleness off the backs of flesh-and-blood machines called slaves or serfs or peasants. When such a situation was combined with a high culture, however, aristocrats used their leisure to become educated in literature, the arts, and philosophy. Such studies were not useful for work, but they occupied the mind, made for interesting conversation and an enjoyable life.

These were the liberal arts, arts for free men who didn't have to work with their hands. And these were considered higher and more satisfying than the mechanical arts, which were merely materially useful.

Perhaps, then, the future will see a world aristocracy supported by the only slaves that can humanely serve in such a post — sophisticated machines. And there will be an infinitely newer and broader liberal arts program, taught by the teaching machines, from which each person could choose.

Some might choose computer technology or fusion engineering or lunar mining or any of the professions that would seem vital to the proper functioning of the world. Why not? Such professions, placing demands on human imagination and skill, would be very attractive to many, and there will surely be enough who will be voluntarily drawn to these occupations to fill them adequately.

But to most people the field of choice might be far less cosmic. It might be stamp collecting, pottery, ornamental painting, cooking, dramatics, or whatever. Every field will be an elective, and the only guide will be "whatever you wish."

Each person, guided by teaching machines sophisticated enough to offer a wide sampling of human activities, can then choose what he or she can best and most willingly do.

Is the individual person wise enough to know what he or she can best do? — Why not? Who else can know? And what can a person do best except that which he or she wants to do most?

Won't people choose to do nothing? Sleep their lives away?

If that's what they want, why not? — Except that I have a feeling they won't. Doing nothing is hard work, and, it seems to me, would be indulged in only by those who have never had the opportunity to evolve out of themselves something more interesting and, therefore, easier to do.

In a properly automated and educated world, then, machines may prove to be the true humanizing influence. It may be that machines will do the work that makes life possible and that human beings will do all the other things that make life pleasant and worthwhile.

26 THE FRIENDS WE MAKE

THE TERM "robot" dates back only sixty years. It was invented by the Czech playwright, Karel Capek, in his play, R.U.R., and is a Czech word meaning worker.

The *idea*, however, is far older. It is as old as man's longing for a servant as smart as a human being, but far stronger, and incapable of growing weary, bored, or dissatisfied. In the Greek myths, the god of the forge, Hesphaistos, had two golden girls — as bright and alive as flesh-and-blood girls — to help him. And the island of Crete was guarded, in the myths, by a bronze giant named Talos, who circled its shores perpetually and tirelessly, watching for intruders.

Are robots possible, though? And if they are, are they desirable?

Mechanical devices with gears and springs and ratchets could certainly make manlike devices perform manlike actions, but the essence of a successful robot is to have it *think* — and think well enough to perform useful functions without being continually supervised.

But thinking takes a brain. The human brain is made up of microscopic neurons, each of which has an extraordinarily complex substructure. There are 10 billion neurons in the brain and 90 billion supporting cells, all hooked together in a very intricate pattern. How can anything like that be duplicated by some man-made device in a robot?

It wasn't until the invention of the electronic computer thirty-five years ago that such a thing became conceivable. Since its birth, the electronic computer has grown ever more compact, and each year it becomes possible to pack more and more information into less and less volume.

In a few decades, might not enough versatility to direct a robot be packed into a volume the size of the human brain? Such a computer would not have to be as advanced as the human brain, but only advanced enough to guide the actions of a robot designed, let us say, to vacuum rugs, to run a hydraulic press, to survey the lunar surface.

A robot would, of course, have to include a self-contained energy source; we couldn't expect it to be forever plugged into a wall socket. This, however, can be handled. A battery that needs periodic charging is not so different from a living body that needs periodic feeding.

But why bother with a humanoid shape? Would it not

be more sensible to devise a specialized machine to perform a particular task without asking it to take on all the inefficiencies involved in arms, legs, and torso? Suppose you design a robot that can hold a finger in a furnace to test its temperature and turn the heating unit on and off to maintain that temperature nearly constant. Surely a simple thermostat made of a bimetallic strip will do the job as well.

Consider, though, that over the thousands of years of man's civilization, we have built a technology geared to the human shape. Products for humans' use are designed in size and form to accommodate the human body — how it bends and how long, wide, and heavy the various bending parts are. Machines are designed to fit the human reach and the width and position of human fingers.

We have only to consider the problems of human beings who happen to be a little taller or shorter than the norm — or even just left-handed — to see how important a good fit into our technology is.

If we want a directing device then, one that can make use of human tools and machines, and that can fit into the technology, we would find it useful to make that device in the human shape, with all the bends and turns of which the human body is capable. Nor would we want it to be too heavy or too abnormally proportioned. Average in all respects would be best.

Then too, we relate to all nonhuman things by finding, or inventing, something human about them. We attribute human characteristics to our pets, and even to our automobiles. We personify nature and all the products of nature and, in earlier times, made human-shaped gods and goddesses out of them.

Surely, if we are to take on thinking partners — or, at the least, thinking servants — in the form of machines, we will be more comfortable with them, and we will relate to them more easily, if they are shaped like humans.

It will be easier to be friends with human-shaped robots than with specialized machines of unrecognizable shape. And I sometimes think that, in the desperate straits of humanity today, we would be grateful to have nonhuman friends, even if they are only friends we build ourselves.

27 OUR INTELLIGENT TOOLS

ROBOTS don't have to be very intelligent to be intelligent enough. If a robot can follow simple orders and do the housework, or run simple machines in a cut-and-dried, repetitive way, we would be perfectly satisfied.

Constructing a robot is hard because you must fit a very compact computer inside its skull, if it is to have a vaguely human shape. Making a sufficiently complex computer as compact as the human brain is also hard.

But robots aside, why bother making a computer that compact? The units that make up a computer have been getting smaller and smaller, to be sure — from vacuum tubes to transistors to tiny integrated circuits and silicon chips. Suppose that, in addition to making the units smaller, we also make the whole structure bigger.

A brain that gets too large would eventually begin to lose efficiency because nerve impulses don't travel very quickly. Even the speediest nerve impulses travel at only about 3.75 miles a minute. A nerve impulse can flash from one end of the brain to the other in one four-hundred-fortieth of a second, but a brain 9 miles long, if we could imagine one, would require 2.4 minutes for a nerve impulse to travel its length. The added complexity made possible by the enormous size would fall apart simply because of the long wait for information to be moved and processed within it.

Computers, however, use electric impulses that travel at more than 11 million miles per minute. A computer 400 miles wide would still flash electric impulses from end to end in about one four-hundred-fortieth of a second. In that respect, at least, a computer of that asteroidal size could still process information as quickly as the human brain could.

If, therefore, we imagine computers being manufactured with finer and finer components, more and more intricately interrelated, and *also* imagine those same computers becoming larger and larger, might it not be that the computers would eventually become capable of doing all the things a human brain can do?

Is there a theoretical limit to how intelligent a computer can become?

I've never heard of any. It seems to me that each time we learn to pack more complexity into a given volume, the computer can do more. Each time we make a computer larger, while keeping each portion as densely complex as before, the computer can do more.

Eventually, if we learn how to make a computer sufficiently complex *and* sufficiently large, why should it not achieve a human intelligence?

Some people are sure to be disbelieving and say, "But how can a computer possibly produce a great symphony, a great work of art, a great new scientific theory?"

The retort I am usually tempted to make to this question is, "Can you?" But, of course, even if the questioner is ordinary, there are extraordinary people who are geniuses. They attain genius, however, only because atoms and molecules within their brains are arranged in some complex order. There's nothing in their brains *but* atoms and molecules. If we arrange atoms and molecules in some complex order in a computer, the products of genius should be possible to it; and if the individual parts are not as tiny and delicate as those of the brain, we compensate by making the computer larger.

Some people may say, "But computers can only do what they're programmed to do."

The answer to that is, "True. But brains can do only what they're programmed to do — by their genes. Part of the brain's programming is the ability to learn, and that will be part of a complex computer's programming."

In fact, if a computer can be built to be as intelligent as a human being, why can't it be made *more* intelligent as well?

Why not, indeed? Maybe that's what evolution is all about. Over the space of three billion years, hit-and-miss development of atoms and molecules has finally produced, through glacially slow improvement, a species intelligent enough to take the next step in a matter of centuries, or even decades. Then things will *really* move.

But if computers become more intelligent than human beings, might they not replace us? Well, shouldn't they? They may be as kind as they are intelligent and just let us dwindle by attrition. They might keep some of us as pets, or on reservations.

Then too, consider what we're doing to ourselves right now — to all living things and to the very planet we live on. Maybe it is *time* we were replaced. Maybe the real danger is that computers won't be developed to the point of replacing us fast enough.

Think about it!

28 THE LAWS OF ROBOTICS

It isn't easy to think about computers without wondering if they will ever "take over."

Will they replace us, make us obsolete, and get rid of us the way we got rid of spears and tinderboxes?

If we imagine computerlike brains inside the metal imitations of human beings that we call robots, the fear is even more direct. Robots look so much like human beings that their very appearance may give them rebellious ideas.

This problem faced the world of science fiction in the 1920s and 1930s, and many were the cautionary tales written of robots that were built and then turned on their creators and destroyed them.

When I was a young man I grew tired of that caution, for it seemed to me that a robot was a machine and that human beings were constantly building machines. Since all machines are dangerous, one way or another, human beings built safeguards into them.

In 1939, therefore, I began to write a series of stories in which robots were presented sympathetically, as machines that were carefully designed to perform given tasks, with ample safeguards built into them to make them benign.

In a story I wrote in October 1941, I finally presented the safeguards in the specific form of "The Three Laws of Robotics." (I invented the word *robotics*, which had never been used before.)

Here they are:

1. A robot may not injure a human being or, through inaction, allow a human being to come to harm.

2. A robot must obey the orders given it by human beings

except where those orders would conflict with the First Law.

3. A robot must protect its own existence except where such protection would conflict with the First or Second Law.

These laws were programmed into the computerized brain of the robot, and the numerous stories I wrote about robots took them into account. Indeed, these laws proved so popular with the readers and made so much sense that other science fiction writers began to use them (without ever quoting them directly — only I may do that), and all the old stories of robots destroying their creators died out.

Ah, but that's science fiction. What about the work really being done now on computers and on artificial intelligence? When machines are built that begin to have an intelligence of their own, will something like the Three Laws of Robotics be built into them?

Of course they will, assuming the computer designers have the least bit of intelligence. What's more, the safeguards will not merely be *like* the Three Laws of Robotics; they will *be* the Three Laws of Robotics.

I did not realize, at the time I constructed those laws, that humanity has been using them since the dawn of time. Just think of them as "The Three Laws of Tools," and this is the way they would read:

1. A tool must be safe to use.

(Obviously! Knives have handles and swords have hilts. Any tool that is sure to harm the user, provided the user is aware, will never be used routinely whatever its other qualifications.)

2. A tool must perform its function, provided it does so safely.

3. A tool must remain intact during use unless its destruction is required for safety or unless its destruction is part of its function.

No one ever cites these Three Laws of Tools because

they are taken for granted by everyone. Each law, were it quoted, would be sure to be greeted by a chorus of "Well, of course!"

Compare the Three Laws of Tools, then, with the Three Laws of Robotics, law by law, and you will see that they correspond exactly. And why not, since the robot or, if you will, the computer, is a human tool?

But are safeguards sufficient? Consider the effort that is put into making the automobile safe — yet automobiles still kill 50,000 Americans a year. Consider the effort that is put into making banks secure — yet there are still bank robberies in a steady drumroll. Consider the effort that is put into making computer programs secure — yet there is the growing danger of computer fraud.

Computers, however, if they get intelligent enough to "take over," may also be intelligent enough no longer to require the Three Laws. They may, of their own benevolence, take care of us and guard us from harm.

Some of you may argue, though, that we're not children and that it would destroy the very essence of our humanity to be guarded.

Really? Look at the world today and the world in the past and ask yourself if we're not children — and destructive children at that — and if we don't need to be guarded in our own interest.

If we demand to be treated as adults, shouldn't we act like adults? And when do we intend to start?

29 GOING BACK TO COAL

THE INCIDENT at the Three Mile Island nuclear plant stirred a great many people to demand an end to nuclear energy. Without it, what will the United States do for energy when shortages of gasoline and fuel oil loom, even in the absence of a war or an embargo?

The answer most often heard is coal. Coal is a far more common substance than oil is, and the United States is uncommonly rich in it. More coal-energy lies under the soil of the United States than oil-energy under the soil of the Middle East. Coal could make us energy-independent for many centuries.

Happy ending? Not quite.

It is hard to get coal out of the ground. Ask any coal miner. Coal mining is, in fact, one of the most dangerous and unpleasant jobs in existence. It involves death by burial underground, and death by lung disease aboveground.

To replace the oil we now use we would have to increase our production of coal fivefold at the very least, but we don't have the coal miners to do it. We could move to more automated procedures, but replacing men by complex machinery is not an overnight job. If we avoid ordinary digging, strip-mining instead, we destroy the land. What's more, however we dig up coal, we spread pollution in the nearby waters.

No matter what we do, it will be expensive. Paying enough people to do the work, or setting up automated machinery, or restoring the land destroyed by strip-mining, or cleaning up pollution will take so much time, effort, and money that, no matter how rich in coal the United States may be, it will never represent cheap energy.

Then too, coal must be transported in vast quantities from the mines to the various furnaces where it is to be burned, and that involves rebuilding our decayed railroad system — a lengthy and expensive job.

Coal might be treated at the mine to convert it to liquid fuel, which is easier to transport and use, but that will mean developing technology and using more water than is easily available in the western coal-rich states.

If all the problems were solved, the coal must still be burned, and that raises the possibility of serious air pollution that could cause far more lung cancer than that caused by radiation leakage from such incidents as Three Mile Island.

Even if all pollution-producing ingredients were removed from coal (an expensive procedure), its burning would produce carbon dioxide. That much is unavoidable, since coal is almost pure carbon.

There is carbon dioxide in the atmosphere already, of course; 0.029 percent as of the year 1900. Burning fuel (wood, coal, gas, oil) has raised that quantity by a tiny bit each succeeding year. Atmospheric content of carbon dioxide is now about 0.032 percent, and by 2000 it will have reached 0.038 percent.

This seems trifling, but carbon dioxide is a heat trap. It lets sunlight through in the daytime, but at night, when the Earth re-radiates its heat as infrared radiation, carbon dioxide absorbs and retains some of it. The Earth is therefore a bit warmer than it would be if the atmosphere lacked carbon dioxide. (Water vapor also acts as a heat trap.)

As the carbon dioxide content of the atmosphere goes up, the Earth will grow a bit warmer, which is known as the greenhouse effect. That extra bit of warmth may be unnoticeable, but it could well shift the balance of freezing/

melting in the polar region just enough to trigger a melting of the polar ice caps year by year, so that the sea level would slowly rise and the coastal rims of the continents would slowly be submerged.

It could be even worse than that. Gases are uniformly less soluble in warm water than in cold. As the temperature goes up, the ocean will grow slightly warmer, and some of the carbon dioxide dissolved in it would be released. That would further increase the carbon dioxide content of the atmosphere. In addition, the ocean would evaporate somewhat more, increasing the overall water vapor content of the Earth's atmosphere.

The additional carbon dioxide and water vapor together would increase the efficiency of the heat trap and raise the temperature of the Earth still further, which would warm the oceans further — and so on. In this way, a "runaway" greenhouse effect could be initiated that might render Earth as hot and as uninhabitable as Venus.

Things may never go to that extreme, but can we take the chance?

We cannot go backward, in other words. We must hang on to nuclear energy, make it as safe as possible, and move forward to safer sources — fusion and solar power. Meanwhile we must limit our energy needs by controlling waste, luxury use, and population

---

30 THE FINAL FUEL

SINCE HUMAN BEINGS discovered how to tame fire many thousands of years ago, we have been burning things for energy — wood, fat, wax, coal, oil, even animal wastes — and we're running low. It may become more and more of a scramble to find enough material to burn for our energy needs.

We could be heading in the direction of energy without fuel, however: energy from wind and water and sun, which don't pollute or run out.

But how does that help transportation, for instance? You can't drive an automobile by putting a sail on its roof or by focusing sunlight on its rear or by setting up a small waterfall under its hood.

You won't have to, naturally. The nonfuel sources of energy can be used to generate electricity; the electricity can charge storage batteries; the storage batteries can run electric cars.

Electric cars are relatively noiseless and nonpolluting, but they tend to be slower than gas-fueled cars. Besides, a fully charged battery won't take you as far as a full gas tank, and it takes considerably longer to charge a battery than to fill a tank.

But then, what's our hurry? It might be better if we didn't travel as far or as fast.

But then we can't electrify everything. It is difficult to imagine electric ships, and just about impossible to imagine electric airplanes and rockets. It would be nice to have fuel available in the future for uses for which it is particularly convenient, or even indispensable.

As it happens, there's a fuel that will never run out and that is, in some ways, the best of all: hydrogen. A given weight of hydrogen will yield three times as much energy as the same weight of gasoline and four times as much as the same weight of coal.

There's an easy source of hydrogen too. Once we have a plentiful and forever-unfailing source of electricity from wind, water, and sun (and from nuclear fusion too, perhaps), that electricity can be used to break up the water molecule into its two elements — hydrogen and oxygen — a technique that has been known since 1800. We have, of course, three hundred million cubic miles of water in the ocean to serve as raw material.

We can allow the oxygen to escape into the air, while hydrogen gas is piped to wherever it is needed through the network we have developed to handle natural gas.

When hydrogen is burned to produce energy, it turns back into water and nothing else. What's more, the oxygen it consumes in burning is exactly equal in volume to the oxygen obtained by breaking up the water molecule, which was released into the air.

Hydrogen is a very light gas that takes up a great deal of room. It is difficult to compress, very difficult to liquefy. How could it be handled for small-scale purposes?

One recent discovery is that an iron-titanium alloy, when cold, can absorb hydrogen in great quantities; when it is heated moderately, the alloy will release the hydrogen again. Imagine a "gas tank" filled with a spongy alloy into which hydrogen can be fed, under pressure, and that will then feed that hydrogen into an engine in small quantities.

There is a serious catch. Hydrogen burns *too* easily. It is, in fact, explosive, and the smallest spark will set it off. (Remember the Hindenburg, that giant dirigible that went down in flames in a matter of a few minutes?)

There is, however, another way to store hydrogen. It can be combined with carbon dioxide (an easily obtained substance) to produce such things as methyl alcohol and methane, fuels that deliver less energy, weight for weight, than hydrogen, but are less explosive.

For that matter, given a convenient source of plentiful energy, we can begin with hydrogen and carbon dioxide and, after a number of chemical manipulations, end up with gasoline, whose molecules are made up of chains of seven or eight carbon atoms with hydrogen atoms attached. Oxygen is again left over and discharged into the air.

We'll have gasoline after all — and forever. What's more, it will be nonpolluting. There will be nothing in it but carbon and hydrogen so that when it burns, we get back the carbon dioxide and water we started with, and nothing else. Nothing gets used up except energy from wind, water, sun, and possibly nuclear fusion, all of which will last as long as the Earth will.

31 THE OCEAN DIFFERENCE

IT IS sometimes advanced as a comforting thought, in the midst of our energy crisis, that there is energy everywhere. Everything, even matter itself, is a form of energy.

However, energy can't be put to use if its concentration is everywhere the same. To use energy, to turn it into work, you have to find it in greater concentration in one place, in lesser concentration in another. You then use the difference.

For instance, where heat energy is concerned, the difference in concentration is represented by temperature. If the Sun shone down on all parts of the Earth equally all the time, then all parts of Earth's surface would be at the same temperature and we could get no work out of it.

The dayside of the Earth is, however, warmed to a greater extent than the nightside, and the tropics are warmed to a greater extent than the polar regions. This means that warm air rises in some places while cold air sinks in others; and warm water rises in some places while cold water sinks in others. This sets up air currents (wind) and ocean currents. Warm water evaporates readily, and the water vapor is carried by the wind to cold regions where it condenses, causing rain and snow.

We can use the energy of the wind, of ocean currents, of moving streams (fed by the rain), because all that energy arises out of differences in temperature at different parts of Earth's surface, resulting from the unequal heating of that surface by the Sun.

Is there any way, though, in which we can cut out the middleman and make use of the temperature difference directly?

It doesn't seem easy at first glance. Nature, working with the planet as a whole, can use differences in temperature on the dayside and nightside, in the tropics, and near the poles. It can span thousands of miles in setting up its "heat engine."

We are limited to smaller distances, at least for now.

Fortunately, we can find a difference in temperature in the oceans over a much shorter span. The ocean surface in the tropics and subtropics, lying under a blazing sun day after day, is lukewarm. But the ocean depths, where the Sun's rays do not reach, are cold. Deep down the ocean water everywhere, even in the tropics, is just a degree or two above freezing.

Nor does the warmth of the ocean surface reach down very far. The surface warmth goes into evaporating water and leaks downward only in enough quantity to keep the ocean from freezing.

We can, in theory, use this ocean difference. Imagine a pipe a little over a thousand feet long anchored vertically in some shallow part of the ocean. Warm water enters a reservoir at the top, and cold water enters a reservoir at the bottom.

The warm water tends to evaporate at the top, and the cold water condenses the vapor at the bottom. A tiny wind is created that runs from the top of the pipe to the bottom.

If there were only a fixed amount of warm water on the top and a fixed amount of cold water at the bottom, the warm water would cool as it evaporated, and the cold water would warm as the vapor condensed. When both were at the same temperature the wind would stop. However, the warm water reservoir is connected to the rest of the ocean, and so is the cold water reservoir. Sunlight keeps the upper layers warm, and the ocean circulation keeps the bottom layers cold, and the slight wind blows forever.

If you have a turbine in the pipe, the wind will turn the turbine and create an electric current.

In this way, you can produce electricity without burning fuel and without producing pollution. You use only sunlight and ocean water. Large numbers of pipes can be set up here and there in the ocean, all producing electricity and going a considerable way toward solving our energy problems.

About the only serious catch (once the engineering bugs are ironed out) are the storms that roil the ocean surface now and then. They roil *only* the surface, however; the depths remain calm under the strongest hurricane. It would then only be necessary to plug the pipe, tighten the anchors, and pull the entire apparatus a hundred feet or so down to wait out the storm. No storm should affect more than a small portion of the pipes once they are spread sufficiently widely over the ocean, and the overall energy supply will not be seriously affected.

32 ENERGY WITHOUT GEOGRAPHY!

IN THE days following World War II, many people viewed uranium as just another fuel that could be exploded in anger or consumed more slowly in calculation — doing more damage than other fuels in the first case, but more good in the second.

Unfortunately, uranium is more than just another fuel — and it is worse.

What the general public didn't realize, at first, was that the exploded atomic bomb did not do all its damage at the moment of explosion. When uranium undergoes fission, the ash it produces is composed of fiercely radioactive atoms, and if these spread out over the land they will carry a deadliness that would last for decades.

Uranium undergoing controlled fission has, in the decades since, produced useful energy for humankind — but the radioactive ash is still produced; it is still there. It must be disposed of safely and must not be allowed to escape into the environment either before or after disposal. This aspect of fission energy makes it a most uneasy substitute for the fossil fuels — coal and oil.

In fact, if we look toward the future we find that coal, oil, and uranium are alike in the respect that we can depend on none of them for the twenty-first century. The oil, for the most part, will be gone by then. The coal, in the quantities needed, will destroy the land in the course of being dug out and pollute the air in the course of being burned. As for uranium, that will have within it, always, the fearsome danger of the radioactive contamination of the world.

Yet we must have energy if civilization is to survive. Where will it come from if oil, coal, and uranium are eliminated as possible sources?

One thing to remember is that we use many sources of energy; we have used them in the past and can use more of them in the future. There is still the energy of the wind, of running water, of the tides, of the Earth's internal heat.

Even all these together may not be enough to supply all of humankind's energy needs. Properly exploited, however, they can offer us a large percentage of the energy we need and relieve the weight of our dependence on oil, coal, and uranium until the really suitable source is developed.

Each of these sources, moreover, is inexhaustible. Wind and running water will last as long as the Sun shines, the tides will be there as long as the Earth turns, and the Earth will have internal heat as long as it exists. All that energy is there and is being expended; it is only necessary to use it rather than letting it go to waste.

But if the wind and running water derive their energy from the Sun, why not turn to the Sun directly? The sunlight that bathes the Earth delivers energy in hours that would last humankind for years.

To be sure, the energy of the Sun is used in vast quantities by plant life to produce the food and oxygen on which animal life depends, and the energy serves to evaporate the ocean and produce the clouds and rain that make all land life possible — so we can't say that the energy of the Sun is going to waste. Still, large quantities of sunlight bathe desert areas where no plants grow and no water evaporates. The sunlight merely turns into heat. Why can we not put that energy to use before its inevitable end as heat?

One way to do this would be to have sunlight bathe large arrays of solar batteries. Each such battery is a wafer of the semimetal silicon to which has been added small amounts of certain impurities. When exposed to sunlight, these wafers will produce electricity.

Such electricity is produced without pollution of any kind and would continue to flow as long as the Sun shone. What a vision of free energy!

But the enormous energy in sunlight is very dilute; it is spread out thinly over large areas. Even if sunlight were converted to electricity with 100 percent efficiency, 3000 square miles of desert area would have to be coated with batteries to serve the world's energy needs. At 10 percent efficiency, a more likely figure of 30,000 square miles would be needed.

To establish and maintain such vast arrays of batteries would require a large capital outlay and, further, would involve considerable running expenditures in repairs, replacement, and general patrolling. The energy would not be entirely free.

What's more, there is the matter of geography. An energy source, however great, however nonpolluting, however permanent, becomes a problem if it divides the world too sharply into energy-producers and energy-consumers. We see this in connection with oil, which some nations have, and other nations need, and we are witnessing the kind of economic hostility to which that can give rise.

Almost all energy sources are geographically concentrated. Some nations are rich in water power, others poor. There are windy areas and calm areas, places with high tides and places with scarcely any, regions where there are easily tapped sources of internal heat and regions with none.

Even sunlight isn't spread out evenly. Where people are most concentrated is where the rains make agriculture most successful and, therefore, where clouds most frequently block the Sun. The underpopulated desert areas collect the energy of the Sun, but the heavily populated nasty-weather areas need it most.

Suppose, though, we could find a source of energy as vast as that of the Sun, one that will also be less dilute and, most of all, less geographically lopsided. Such an energy source does exist in the form of nuclear fusion. Not fission — fusion!

The two are quite different. In nuclear fission, very large atoms are broken into somewhat smaller ones. In nuclear fusion, very small atoms are combined into somewhat larger ones.

In nuclear fission, the chief energy source is uranium, a rather rare metal, which exists in useful quantities in only a few places in the world. In nuclear fusion, the chief energy source is a form of hydrogen called deuterium, which is found wherever water exists.

In nuclear fission, large atoms are broken into very dangerous radioactive atoms. In nuclear fusion, small atoms are combined into helium, the safest substance known.

In nuclear fission, large quantities of uranium must be used. If something goes wrong and the process gets out of hand, the uranium could melt and radioactivity could escape into the environment. In nuclear fusion, tiny quantities of deuterium will be used at one time. If anything goes wrong, the process just stops.

In nuclear fission, considerable energy is produced per pound of fuel. In nuclear fusion, four times as much energy is produced per pound of fuel as in the case of fission.

It would seem, then, that every possible factor favors nuclear fusion — but there are some drawbacks too.

First, while helium is the chief product in fusion, some quantities of mildly radioactive materials are also produced. These will have to be kept out of the environment, but the problem is expected to be much less difficult than in the case of fission.

Second, and more annoying, scientists have not yet succeeded in setting off a controlled fusion reaction.

The trouble is that to get the small atoms to smash together with sufficient force to allow them to coalesce and fuse into larger atoms, the temperature of those atoms must be raised to more than 100,000,000° C. High temperatures for the purpose are to be found in the center of the Sun, which runs on the energy of nuclear fusion — but the enormous gravity of the Sun can hold such unimaginably hot material in place while it is fusing. What can it do on Earth?

For three decades, American, British, and Soviet scientists have been trying to design and produce strong magnetic fields that will hold electrically charged atom fragments in place while the temperature is raised to the necessary high figure to start fusion going. Some of the most advanced devices for producing such magnetic fields are to be found at Princeton University. Scientists have come steadily closer to their goal in this respect but even now are not quite there.

Another attack on the problem has been to use a laser beam. Lasers can concentrate large amounts of energy into a tiny area. If that energy is concentrated on a quantity of deuterium, the deuterium may be heated to a high enough temperature to begin fusing before its atoms have a chance to move out of the way of the laser beam.

What we need right now are magnetic fields that are just a little stronger, a little more subtly designed perhaps. Or else laser beams that are a little stronger, a little more tightly focused, a little better adapted for the specific purpose of initiating fusion.

Scientists may be on the edge of success in each of these two directions.

Even after the solution is arrived at, however, it may well take thirty years to solve just the engineering problems of setting up large power stations, of designing methods for keeping a constant source of fuel moving into the reactor, for maintaining the lasers in action or the magnetic fields in being, and for guarding against radioactive pollution.

If all goes well, the twenty-first century could see the world moving into a major reliance on fusion power and into a new period of cheap energy.

From nuclear fusion, we can produce electricity more or less directly. We can also use the energy to combine water and carbon dioxide to produce oxygen, and the carbon-hydrogen compound, methane, which is the chief constituent of natural gas. From methane we can build up more complicated carbon-hydrogen molecules like those in gasoline.

In the twenty-first century, then, we could still be burning gas and oil — but it will be gas and oil without the impurities that produce air pollution. It will be gas and oil that cannot be used up, for there are enough fusible atoms in the oceans to last humankind for many millions of years.

Moreover, the prime source of the energy-yielding atoms will be the world's oceans, which means that nuclear power stations can be located anywhere along the continental coastlines. There are virtually no nations and regions without access to such coastlines, so it would be, for the first time in man's history, a case of energy without geography.

As far as energy is concerned, there will no longer be any division into haves and have-nots. If we can learn to control population and avoid war, then, once fusion power becomes a reality, all the world will be haves.

33 CONVERTING IT ALL

ENERGY can be produced at the expense of matter and in very large quantities. If a single pound of matter were converted entirely into energy, it would produce an amount equivalent to the muscular energy expended in an entire day by the human population of all the Earth.

Let's put it another way. We know the awesome power of the hydrogen bomb — the result of converting 0.7 percent of the hydrogen used as nuclear explosive into energy. If *all* the hydrogen were converted into energy, a hydrogen bomb of a given size would be 140 times as destructive as it is.

Nuclear engineers are trying to bring the hydrogen fusion reaction under control, to learn to make it progress under controlled conditions. Controlled hydrogen fusion would give us a source of energy so rich that it would last as long as the Earth is likely to.

Yet that is only one one-hundred-fortieth of the energy that's really there. Is there any way we can suck *all* the energy out of matter?

In theory, there is. The common particles that make up matter are electrons, protons, and neutrons. Each one has an antiparticle, a kind of mirror image that has one key property opposite in nature to that of the particle itself. In other words there are antielectrons, antiprotons, and antineutrons, out of which antimatter can be built up.

Whenever a particle meets its own antiparticle, the two cancel each other and undergo "mutual annihilation." Nothing but energy, equivalent to that of the amount of mass that has disappeared, is left behind.

An ounce of matter and an ounce of antimatter, brought together, would explode at once, forming a ball of intense and deadly radiation. However, if the matter and antimatter were in the form of a thin drizzle of antiparticles being joined in a very slow and steady manner, energy might be formed at a manageable rate. Such controlled matter-anti-matter annihilation would be an energy source 140 times as rich as the same amount of matter undergoing controlled hydrogen fusion.

There are catches, of course. For one thing, antimatter, or even individual antiparticles, does not exist in nature except for small quantities in cosmic rays or in certain radio-active transformations. Antiparticles can be formed in the laboratory but, again, in very small quantities. What's more, the energy required to form the antiparticles is considerably greater than the energy we would get back from it if we allowed it to combine with particles.

Second, even if we could form antiparticles in quantity, where would we keep them? The world is made up entirely of ordinary particles, and any antiparticles would have to be kept safely penned up, possibly in a tight electromagnetic field, out of all contact with ordinary particles, till we wish to begin the controlled interaction. The problem of working out the technique would be formidable.

As it happens, protons and neutrons, scientists now think, are made up of still more fundamental particles called quarks and antiquarks that come in a rather large variety of forms. These quarks and antiquarks can combine only in certain combinations to form protons and neutrons — plus hundreds of other particles that don't appear in ordinary matter but can be made to appear under conditions of very high energy concentrations in the laboratory.

We are only on the edge of understanding quarks, antiquarks, and their combinations, but as we learn more and more it is possible we may learn to manipulate them, with ingenuity, in ways not found in nature. (After all, we have learned to outdo nature in our handling of atoms and of subatomic particles.)

Give us time, and we may learn to build synthetic particles from quarks, particles unlike any that occur naturally. The quark particles we know, besides protons or neutrons, last only exceedingly tiny fractions of a second, but what if we can design some that endure for minutes — long enough for them to be put to use. They may even be sufficiently different from any natural quark combination not to react with ordinary matter, so that they may be stored.

And what if we can manufacture artificial particles like those we have already formed, but with an antiquark wherever a quark should be and vice versa. These should have the same stability and inertness as their mirror images.

Then, if these two sets of artificial particles, mirror images of each other, are mixed at a slow and controlled rate, we would have the ultimate fuel, one in which *all* the matter is converted to energy at a rate slow enough to be useful to us.

Of course, it's a wild dream. It's unlikely such artificial particles are possible, or could be formed without impossible energies, or could have the appropriate properties. Yet nuclear power of any kind was an equally wild dream (if anyone could even dream it) in 1880.

34 PACKAGED WATER

OUR PLANET has an overwhelming amount of water. The total water supply of Earth is over two hundred thousand times as great as humanity needs. Why, then, is humanity worried about its water supply? Well, there's a catch.

Fully 97.4 percent of all the water on Earth is the salt water of the ocean, and the only thing human beings can use salt water for is to fish in and to float ships on. For drinking, for washing, for irrigation, for all industrial needs, only fresh water will do, and that makes up only 2.6 percent of the whole.

Even so, the fresh water supply alone is more than five thousand times as great as humanity needs, and that should be enough.

There's another catch. Unfortunately, 98 percent of all the fresh water on Earth is in the form of ice that remains frozen the year round, and can't be used.

What's left, the *liquid* fresh water on Earth, is only about one two-thousandth of all the water there is.

Even that is enough to supply about 100 times humanity's needs, so it would appear that we were still safe.

But then, the liquid fresh water is not distributed evenly on the planet. What's more, the rain and melting snow that replenish it at a rate that makes for a total turnover every five years is also not evenly spread, either in space or in time. There are steaming rain forests in some parts of the world and arid deserts in others; there are disastrous floods at some times and equally disastrous droughts at others.

Consider also that humanity's needs are increasing as its

numbers increase, as it produces more and more manufactured goods, and as it tries to grow more and more food.
(It takes 200 tons of water to make a ton of steel, and 8000
tons of water to grow a ton of wheat.)

To top it all off, we are freely using our lakes and rivers
as sewers and are polluting more and more of our limited
fresh water supply to the point of uselessness. We are therefore in serious danger of running dreadfully short.

What do we do? Naturally, we must limit our population
growth and end our water pollution, but is there any way
we can increase our fresh water supply?

One suggestion is to look to the world's ice, about 8 percent of which is in Greenland. The Greenland ice sheet
(with an area three times that of Texas) is constantly producing icebergs, which are giant lumps of frozen fresh
water that slowly and wastefully melt back into the salt sea.
About 400 icebergs pass Newfoundland and move into the
open ocean each year. Can some of them be trapped, as
neatly packaged water, and brought to places where that
fresh water is desperately needed?

Perhaps, but it is not nearby east North America and
west Europe that need water most. The Pacific coast of the
Americas and the Indian coast of the Middle East are more
dangerously parched, and there is no practical way to drag
the North Atlantic icebergs around South America or Africa
to get them to their thirsty destinations.

But wait. In Antarctica there is an ice sheet ten times the
size of Greenland's, one that has an area one and a half
times that of the whole United States. Flat-topped icebergs
are produced there that are much larger than any that
Greenland's ice sheet can produce. In 1956, a single Antarctic
iceberg was sighted that was 200 miles long and 60 miles
wide — a single piece of free-floating ice with an area half
again that of the state of Massachusetts.

Even a comparatively small Antarctic iceberg would contain enough water, when melted, to supply the water needs of 700,000 people — drinking, washing, irrigation, and industry — for one month.

Imagine such an iceberg dragged slowly northward across the tropic waters directly to the Middle East or to California. The iceberg would have to be trimmed to a shiplike form to reduce water resistance; it would have to be insulated on the sides and bottom to reduce melting; and at destination it would have to be sliced up, melted, and the water stored.

These engineering details may be worked out eventually, and then the world can have its water — on the rocks.

35 THE ENDLESS MINE

ABOUT forty-five centuries ago, human beings discovered scarcity; they discovered what "used up" meant.

Pre–civilized humanity used resources that renewed themselves — wood, grain, meat, water; or resources that seemed endless — flint, clay, sand, sunlight. But then, along the shores of the eastern Mediterranean, people established civilizations and learned how to obtain metal. By 3500 B.C., the deliberate smelting of copper ore was a well-established industry in the Middle East.

Copper, when mixed with tin, forms bronze, which is hard and tough enough to use for tools and weapons. In the Bronze Age that followed, ores that contained copper and tin were avidly sought. The very term "metal" is from a Greek word meaning "to search for."

Copper is a relatively rare metal, but tin is rarer still, and by 2500 B.C., the tin-ore supply of the Middle East gave out entirely. For the first time, human beings faced the final disappearance of a natural resource, as opposed to temporary shortages of food and water.

The disappearance was only local, however. The Earth was large, and there were tin mines elsewhere. The Phoenician navigators scoured the Mediterranean for metal ores; by 1000 B.C. they were venturing as far as the British Isles in their search for tin.

Since those days humankind has been using more and more metals of all kinds, moving on to iron and steel, to alloy steels making use of nickel, chromium, tungsten, molybdenum, vanadium, and others, then to aluminum, magnesium, and titanium.

In the five thousand years and more that human beings have been using metals, the good mines, where the ore is richest, have been inexorably exhausted. To be sure, techniques have been developed for getting metal, profitably, out of thinner and thinner ores, but where will it all end? Are we going to run out of key metals altogether — not just here and there, but everywhere?

No, not necessarily! Metals are elements; they are made up of characteristic atoms which, under earthly conditions, are unchanging and always exist. To be sure, in the process of mining and using metals, we spread them out more thinly and it takes energy to bring those atoms together again. Still, if we must concentrate, it is sometimes easier to collect thinly spread atoms from a liquid than from a solid mixture.

As it happens, atoms of every kind wash down from the land into the ocean, little by little. Some move into the ocean at a faster rate than others, but no atom is completely immune to the washing effect of rain, ground water, and rivers.

As a result, about 3.33 percent of the ocean is solid matter in solution.

Imagine a swimming pool 50 feet long, 30 feet wide, and 6 feet deep that is filled with sea water. Evaporate that quantity of sea water and 9.25 tons of solid matter will be left behind, most of it made up of atoms of sodium and chlorine which, in combination, make up sodium chloride — common salt.

Beyond that, in various combinations, are 750 pounds of magnesium, 500 pounds of sulfur, 230 pounds of calcium, 220 pounds of potassium, 27 pounds of bromine, and about 28 pounds of other things, including traces of copper, silver, gold, uranium, and even radium. The total quantity of some of these rare metals in an ocean large enough to fill 20 million billion swimming pools of the size just mentioned is enormous. The ocean contains 5 billion tons of uranium, for instance, and 8 million tons of gold.

In addition, some metals settle out of solution around a pebble or some other object as a nucleus, forming nodules scattered over the ocean floor. There may be as many as 31,000 tons of such nodules per square mile of the Pacific floor. The nodules are particularly rich in manganese, but contain considerable amounts of nickel, copper, and cobalt as well.

Some of the richer constituents of the ocean solids can be extracted profitably — salt, of course, since ancient days, but modern chemists extract magnesium and bromine too. Where human chemists fail, lower forms of life can succeed. Seaweed, such as kelp, will extract the tiny traces of iodine from sea water. Chemists then burn the kelp and extract iodine from the ash.

It is not difficult to see a future in which metallic nodules are dredged up from the sea bottom and in which more and more elements are extracted from the sea water itself.

Nor need we fear using up this vast mine, for everything we take out of the ocean washes back into it sooner or later.

Again, concentrating a thinly spread out element takes energy. That, however, is just another way of saying that if we had enough energy — solar or fusion or both — we would not have to fear any material shortages, if humankind's numbers and demands remained reasonable.

What's more, there's a second mine on earth that may be as useful as the ocean — and we'll get to that in the next essay.

36 COMPLETING THE CYCLE

On Earth, things move in cycles. That is why, after 4.6 billion years of existence, our planet is still young, active, and full of life.

Water runs from the uplands to the sea, quickly as rivers and seepingly as ground water. It may delay a while and collect in ponds and lakes, but it always ends in the ocean (or, occasionally, in inland seas). It is a natural flow under the pull of gravity.

Why did all the water not end in the sea, leaving the land desert-dry billions of years ago? Because that's only half the cycle.

The other half, powered by the energy of the Sun's radiation, evaporates water from the ocean surface, raises water vapor by the millions of tons a couple of miles into the air, carries it over the land, and drops it as rain.

A second example: Animals convert food and oxygen

into carbon dioxide and water. That's half the cycle. Plants, powered by the light of the Sun, convert carbon dioxide and water into food and oxygen. That's the other half.

There are many other cycles on Earth, most interconnected, some involving life and some not, and in every case there must be an energy input to complete the cycle. Since the Sun is on hand, with copious energy that has lasted for billions of years and will last for billions of additional years, that is no problem.

But now humanity and its technology are on the scene, and in every direction change is carried through half a cycle. Rich pockets of ore are smelted for metals; the metals are used in devices; the devices are finally discarded and allowed to rust or just to lie there. Trees are chopped down; wood is converted to paper and a myriad of paper products; the paper products are used and thrown away.

In short, technology seems to be converting natural resources into trash at an enormous rate. Some of the trash can be recycled by natural processes, but not at a rate equal to that of its production. Some trash consists of new products invented by human beings, which can't be recycled at all by natural processes.

The result is that one of the great problems of an industrial society is disposing of the mountains of trash produced each day. Humanity is in serious danger of drowning in its own waste.

Yet everything is composed of atoms, and atoms (barring nuclear processes, which take place only rarely) are eternal and unchangeable. The atoms present in the useful product are also present in the discarded waste.

Why should not the various atoms in the endless trash we turn out be rescued, brought back, and put to use again? Why should not our trash heaps and garbage dumps, which are such eyesores, such breeding grounds for disease, such enormous and nonproductive expenses, serve as *another* end-less mine like the ocean?

The answer to this, as to almost everything, involves energy. From useful product to waste is the easy half of the cycle, the downhill-and-spontaneous-change half. To bring about the other, *uphill*, half of the cycle, from waste back to useful product, requires an input of energy, both human and inanimate. So far it still takes less energy to continue to despoil the pockets of Earth's resources and to continue to produce waste than to tackle the waste itself and complete the cycle.

This may change for two reasons. First, the steady decline of old resources will increasingly force us to turn in desperation to recycling our wastes. Second, new energy sources should yield new tools with which to complete the cycle.

If we develop controlled nuclear fusion, for instance, we may be able to devise a "plasma torch," a fusion-powered spurt of ultrahot gas that attains temperatures of hundreds of thousands of degrees.

In the heat of such a torch, all molecules, without exception, would be broken into the individual atoms that make them up. The myriad of different wastes we produce, ladeled endlessly into one end of a plasma-torch incinerating plant would be converted into a jumble of atoms (of which a dozen varieties would make up the bulk).

These atoms, as they cool, would be separated by appropriate processes into metals and other solids, into gases and liquids. The end products of the plasma-torch incinerating plant would be, for the most part, elements, element mixtures, and simple compounds — all of which could easily be used by technology to re-produce the useful products from which they came.

Human technology would imitate nature, then, by dealing with complete cycles (at the expense of energy) rather than with half-cycles, and ordinary mining will become less and less necessary. We can use and re-use the materials we already have and rarely, if ever, will we need to supplement them with anything new.

37 DEEP, DEEP, DEEP!

THERE IS, alas, an exception to the plasma torch and its potential capacity for converting everything into its constituent atoms; for demolishing all wastes, all trash, all polluting materials, and reducing them to elements and simple compounds that can be recycled in the service of humanity.

One polluting material — the radioactive ash produced by nuclear reactors — remains immune to even the plasma torch, which can break molecules into separate atoms but cannot reach the atomic interior where radioactivities lie. Such ash is now being produced in sizable and dangerous quantities. What's more, if our civilization survives and grows to depend more and more on fission energy, the quantities produced will increase.

The radioactive atoms must be kept out of the environment, for they are deadly. In some cases, they must be kept out of the environment for thousands of years before the time is reached when the atoms have broken down and become safe.

Can they be concentrated in solution, stored in tight, stainless steel containers, and buried? Are we sure they won't leak?

Can they be fused in glass and placed on the snow wastes

of Antarctica, or in salt mines a mile below the surface, or on the ocean bottom? Can we be sure the glass won't be broken, the radioactive atoms leached out and somehow escape?

Can we fire the ash into outer space? Even supposing we can afford the energy, what happens if a rocket explodes shortly after takeoff and spreads masses of radioactivity through the atmosphere?

In short, we need a foolproof method that will stay fool-proof for thousands of years. Is there such a method? Well, I have thought of something that, as far as I know, has not yet been suggested.

The Earth's crust is not a solid piece, but consists of large separate plates, which fit each other closely, but are in slow independent motion. In some places, the plates move very slowly away from each other, producing a rift in the crust into which hot materials from the Earth layers beneath slowly well upward and solidify. Such a rift runs down the center of the Atlantic Ocean, for instance.

In other places, the plates move very slowly together. They may crumple when they meet, forming vast mountain ranges. Thus, when the plate carrying India moved into the rest of Asia, the Himalaya Mountains were formed. If the plates move together more quickly than that, there is no time for crumpling to take place. One plate then moves under the other, melting in the hot rock layers beneath, dragging the Earth's surface downward and forming the "ocean deeps" which parallel certain island chains, such as the Marianas and the Philippines.

What causes these slow plate movements? The answer is not certainly known, but one rather plausible suggestion is that there are very slow circulatory movements in the hot "mantle" that underlies the crust of the Earth. The hot rock flows very slowly across under the crust, then moves down into the depths, then back in the opposite direction at a great depth, then up again to begin the cycle all over. By friction, the mantle drags the crustal layers above this way and that as the rock slowly moves, taking thousands of years to complete one cycle.

In that case, suppose we reconsider the technique of fusing radioactive wastes into glass and dropping it to the bottom of the ocean. Ordinarily, if it just lies there for thousands of years, it is conceivable that quantities of radioactive material may slowly dissolve out and contaminate the ocean water.

But what if we choose our place carefully? What if we choose a place on the ocean floor where the plate is slowly sinking beneath its neighboring plate? Slowly, a fraction of an inch a year, might the glass prisons of radioactivity come to sink with the plate and be carried into the bowels of the Earth?

Before there is a reasonable chance for the radioactive material to dissolve into the ocean water, the glass will be buried in and surrounded by rock. Once the glass is in the rock, it may very slowly be deformed and shattered. As it sinks lower into steadily higher temperatures, it may melt.

In these cases, however, the radioactive atoms are not liberated, except into hot rock from which they cannot escape. They are lower and lower — deep — deep — deep — then back across the lower depths of the mantle. Eventually, they will be brought upward again, in a huge swirl to one of those places on Earth's surface where plates are moving apart.

But that will be many thousands of years after their disappearance in the ocean deeps, and by then most or all the radioactivity will be gone.

Is this a practical notion for the disposal of enormously dangerous garbage? Perhaps not, for there may be serious catches in it.

In fact, after this essay was first published, I received a

letter from Roger L. Larson of the Lamont-Doherty Geological Observatory of Columbia University. It turned out that my notion of burying wastes in the deep sea trenches had already been thought of — my best ideas are always used up before I can get to them — and it had been decided the method wasn't good enough. The rate of burial would not be fast enough and the material wouldn't be buried in the mantle but scraped off as the plate moves downward. However, the essay apparently prompted Dr. Larson and his colleague, Dr. Walter Alvarez — now at the University of California — to investigate the possibility of burying wastes in deep ocean sediments. They have been designing experiments for testing the procedure, and I hope it works.

---

38 THE VANISHING ELEMENT

UNDER earthly conditions, eighty-one different elements are made up of atoms that are never destroyed. The atoms may mix with each other or combine tightly, but, given energy, they can always be isolated again.

Thus, iron combines with oxygen, rusting in the process, but iron can always be salvaged from the rust, and be none the worse for the experience. In fact, chemists have reached the point where they can arrange and rearrange atoms at will; and the superhot plasma torch, when it is developed, will convert all substances to a mixture of the separate elements and their atoms.

Helium, however, is an important exception, because it will resist the conversion. Helium is composed of the second smallest atoms there are. The helium atom is so small that Earth's gravity cannot hold it efficiently against its own random motions in the atmosphere. Some of the gas always leaks away into outer space and never comes back. Helium, therefore, occurs in the atmosphere in only about 1 part in 1.4 million. Then why is there any helium in the atmosphere at all? Why hasn't it all leaked away? Because additional helium forms slowly in the soil.

Uranium and thorium are not among the eighty-one stable elements. Their heavy atoms break down very slowly into smaller atoms, and those smaller atoms include helium. The helium accumulates in the soil very slowly and is swept out from crannies and crevices by other, more copious gases — notably by accumulations of natural gas.

Natural gas often contains 1 percent or so of helium. In the gas from some wells, the concentration goes up to 8 to 10 percent, and helium can easily be obtained in considerable quantity from such wells.

No matter how helium is used, some of it always leaks away into the atmosphere, where it is spread too thinly to be collected again profitably (even though the total quantity in the atmosphere is 4.5 billion tons). Eventually, it leaks into space.

To be sure there is a still lighter gas — hydrogen — which we use constantly without fear of loss. The reason is that hydrogen atoms combine easily with other atoms to form heavy combinations that do not escape into space. Earth's hydrogen supply exists predominantly in combination with oxygen, as water — and water molecules are safely held by Earth's gravitational field.

Helium atoms, on the other hand, combine with nothing! There is no way to make helium part of a heavy combination that will not escape. It can be confined physically, but such confinement is never permanent or perfect, so helium is constantly being lost.

Are we in danger of running out altogether? Of course we are. The world's entire helium supply is obtained from natural gas, and the natural gas supply of the Earth is impermanent. Our American supply may be gone in twenty years and the world supply in forty.

After that, helium will be rare and precious indeed. It will continue to form in the soil, to be sure, but only very, very slowly. What we have flushed out with our natural gas has been the accumulation, literally, of billions of years of uranium and thorium breakdown. What little helium we will be able to get will have to come out of the thin content in the atmosphere.

Does this loss of our helium supply matter?

Well, a temperature of $-273.16°$ Celsius is absolute zero, the coldest possible temperature. As temperatures sink lower and lower toward this bottom, gas after gas first liquefies and then freezes. By the time a temperature of $-259.14°$ C is reached (just 14 degrees above the absolute) every existing substance, but one, is frozen.

That one exception is helium, which even at 14 degrees above absolute, is still a gas. It doesn't become a liquid until it reaches a temperature of 4 degrees above absolute, and it stays a liquid all the way down to absolute zero. Liquid helium, to freeze, has to be put under pressure.

It is easy to use a liquid/gas combination to reduce temperature. When liquid evaporates into a gas, its temperature drops — a principle used in manufacturing ordinary refrigerators and air conditioners. That means that helium, the only liquid/gas combination that can exist at less than 14 degrees above absolute zero, is the only substance we can use to cool materials down to the immediate vicinity of absolute zero. There is no substitute.

And does *that* matter? We'll take that up in the next essay.

39 THE NEAREST SOURCE

IN THE preceding essay, I talked about the peculiar properties of the element helium, and how it alone remains liquid and unfrozen in the neighborhood of absolute zero. Now let's continue.

Matter exhibits unusual properties in the neighborhood of absolute zero. In 1911 the Dutch physicist, Heike Kamerlingh-Onnes, who had first produced liquid helium three years earlier, checked the electrical resistance of mercury at those very low temperatures.

He expected the resistance to decrease to very low values, for electrical resistance decreases with temperature as a general rule. He did *not* expect that at a temperature of 4.12 degrees above absolute zero the resistance of mercury would vanish *completely*, but that's what it did. If an electric current were produced within a ring of frozen mercury, it would continue to flow forever without diminution as long as the temperature were kept below that figure.

This phenomenon is called superconductivity. A number of other elements and mixtures of elements were found to be superconductive at temperatures in the neighborhood of absolute zero.

Just think of transporting electricity through superconductive wires over long distances. There would be zero loss and much energy would be saved in this way. (We would have to subtract from the energy saving the energy used to maintain low temperatures, of course.)

Furthermore, as more and more electric current is poured into a superconductive substance, stronger and stronger magnets can be built, magnets much stronger than could exist without superconductivity. There are many other technological uses to which superconductivity can be put.

For superconductivity, we need helium, however, and the only practical sources are natural-gas wells. When our natural gas is gone, our helium will be gone. What do we do then if we want to take advantage of superconductivity?

For one thing, scientists have been searching for materials that turn superconductive at unusually high temperatures; some complex metal mixtures turn superconductive at temperatures as high as 21 degrees above absolute zero. At that temperature, hydrogen is liquid, so that we can imagine liquid-hydrogen superconductivity, and we'll never run out of liquid hydrogen.

There will probably never be many substances, however, that will be superconductive in liquid hydrogen, and that will severely limit our options. There are other interesting properties of very low temperatures in the liquid-helium range that simply can't be shifted into the liquid-hydrogen range, and, without helium, much in the way of research and potential advances in technology may become impossible.

Where, then, will we get our helium when our natural supply is exhausted forever?

There is some in our atmosphere, billions of tons actually, but it is very thinly spread out. There may still be some in our soil here and there, or even in the soil of the Moon or Mars, but that too would be very thinly spread out.

For a really solid supply in large and reasonably concentrated quantities we would have to look for a world larger than the Earth, one with a gravity capable of holding the light and fugitive helium atoms.

The nearest large world is the Sun, but it is impractical to think of mining the Sun if there is any other alternative at all.

The nearest *practical* source (using the word *practical* in a very generous way) is Jupiter, a giant planet composed mostly of hydrogen; but helium is the main secondary component.

Getting close to Jupiter is a problem, however, for its huge magnetic field contains deadly concentrations of high-energy particles. It has nine tiny satellites at distances of 7 to 14 million miles from the planet, however, and on these human beings could set up bases. Even the outermost large satellite, Callisto, might be far enough outside the main radiation blanket to serve as a human base.

From Callisto, it may be possible to set up a robot-run, automated base on Jupiter's innermost satellite, Amalthea, which is only 70,000 miles above Jupiter's cloud layer.

We can then imagine sweepers of some sort, launched from Amalthea, passing through the outermost layers of Jovian atmosphere and returning with a cargo of hydrogen and helium.

The two gases could be separated, the hydrogen serving as a source of deuterium for the fusion energy that would run the whole shebang and the helium being fired out in thin containers into orbits that would intersect that of Callisto.

Then, on Callisto, there could be built up the large reserve supply of helium to supply Earth's continuing needs without end.

No, we can't do it now — but fifty years ago we could just barely fly across the Atlantic.

40 THE MIDDLE SEA

THE MEDITERRANEAN SEA (Latin for "in the middle of the land") was the middle sea of antiquity. Western civilization bloomed around its shores three thousand and more years ago.

There is a larger and potentially more important middle sea in the world, which we don't think of because the flat map we usually see distorts it. Look, instead, down at the North Pole on a globe, and you will see that the Arctic Ocean is the true Middle Sea.

The Arctic Ocean is a nearly enclosed and nearly round arm of the sea. On one side is the vastness of the Soviet Union and Scandinavia, with Europe and China beyond. On the other side is the only slightly lesser vastness of Alaska and Canada, with the United States beyond.

Bordering the Arctic Ocean are millions of square miles of tundra that are too cold for much human habitation and have the further disadvantage that the sea that should connect them is permanently ice locked.

Does the Arctic Ocean have to be ice locked? Does the fact that it is at the North Pole make the ice a necessity?

Perhaps not. One of the reasons there is ice in the Arctic is that the ice already there absorbs only 40 percent of the sunlight that falls on it, reflecting the rest. It is not so much that the polar Sun is weak as that so much of its light is uselessly reflected that keeps the ice from melting in the summer.

Without the ice, the bare water of the Arctic would absorb 90 percent of the sunlight falling on it. It would then absorb enough heat to keep it from freezing permanently. The small quantity of ice formed in the winter would melt in the summer. It would seem that only because the Arctic Ocean iced over in the Ice Age does it stay iced over now.

Some years ago, a Soviet geologist suggested that the Arctic ice be covered by coal dust. The black coal dust would absorb heat and melt the ice. The bare water would then continue to absorb heat, and the Arctic Ocean would not refreeze.

It would take an immense quantity of coal dust to cover the millions of square miles of Arctic ice — but perhaps not all of it need be covered. Even if only a sizable fraction of the ice were melted, enough additional sunlight would be absorbed to lower the coal dust requirement for the ice that would be left.

Since the ice floats in water, it displaces an amount of water equal to its weight. As the ice melts, it would merely fill in the hole left by the displaced water, so to speak, and the sea level would not rise one inch.

With the Arctic ice gone, the cold Arctic breezes would be moderated, the long hours of summer sunshine would not be reflected and wasted so much, and the shores of Alaska, Canada, Scandinavia, and Siberia would enjoy milder weather.

Millions of square miles could support many times the number of people they now support, and the whole Northern Hemisphere would have better weather.

How pleasant!

These days, though, we have learned not to trust easy paradises without a closer look. As population surges north on both sides of the world, would the enmity between the Soviet Union and North America become worse? Would Canada, growing stronger, incur the enmity of the United States?

And what of the natural effects? Will a milder Arctic result in the melting of the Greenland ice cap by slow degrees? That ice cap is on land and, as the water of the melting ice flows into the ocean, the sea level will rise. If all the ice in Greenland melts, the sea level will rise twenty feet, displacing millions of people.

Of course, with the Greenland ice gone, the polar weather will be even milder and Greenland itself might support a sizable population. If the change in sea level is slow enough and the nations of the world cooperate, will the final situation represent an overall improvement?

But then, the ice will be truly gone only in the summer. It will snow in the winter. Since water yields more vapor than ice does, a bare Arctic Ocean will result in more precipitation on the surrounding continents than now falls — meaning, in wintertime, more snow.

It could be that more snow would accumulate in winter than would melt in summer, even in the milder polar weather. The accumulated snow would reflect sunlight and make the situation worse. In short, there may be the start of a new Ice Age which will, in the end, freeze over the Arctic Ocean once more, leaving things infinitely worse than they are now.

Perhaps, then, we had better leave the Arctic as nearly alone as we can — till we understand more about the workings of the weather and can better judge the consequences of our well-meant deeds.

41 LUBRICATING THE EARTH

EVERY once in a while the Earth trembles. On the scale of the planet itself, this is a small thing — just a brief, tiny twitch. On the human scale, however, it is enormous; it is the only natural phenomenon that can kill hundreds of thousands of people and destroy billions of dollars' worth of property in five minutes.

We call it an earthquake.

On January 24, 1556, an earthquake struck in the province of Shensi in China and is supposed to have killed 830,000 people. That is still the record for earthquake deaths. On December 30, 1703, an earthquake killed 200,000 people in Tokyo, and on October 11, 1737, another killed 300,000 in Calcutta.

On November 1, 1755, the city of Lisbon was destroyed by an earthquake and by the tsunami (or "tidal wave") that it triggered, and 60,000 died.

As time goes on, earthquakes are bound to get more deadly, simply because there are continually more people on the Earth and because the works of human beings are becoming more numerous, complex, and expensive.

In 1906, for instance, an earthquake destroyed the city of San Francisco, killing 700 people, leaving 250,000 homeless, and doing $500 millions' worth of property damage. If a similar earthquake struck the city today, it is likely more people would die and be left homeless and the amount of property damage would be enormously greater.

What can we do? Can we predict an earthquake so that people can at least get out of the way in time?

We may be able to. Some preliminary changes seem to presage a quake. The ground can hump upward in regions subject to quakes. Rocks can pull apart slightly, causing changes in water level in wells, or in the electrical and magnetic properties of the soil.

We don't detect some preliminary tremblings, but animals living closer to nature do and grow fearful. Horses begin to rear and race, dogs howl, fish leap. Animals like snakes and rats, which usually remain hidden in holes, suddenly come out into the open. In zoos, chimpanzees become restless and spend more time on the ground.

In China, where earthquakes are more common and damaging than in the United States, the people are encouraged to watch for and to report unusual animal behavior, any strange sounds in the Earth, any shifts in the level of well water, or even unusual flaking of paint.

The Chinese report that they have been able to predict some earthquakes and claim to have saved many lives in connection with an earthquake in northeastern China on February 4, 1975. Another, on July 27, 1976, was *not* predicted, however, and a whole city was wiped out.

But the evacuation of a city is troublesome and produces almost as much economic dislocation and personal hardship as an earthquake. Besides, even if people do move, they cannot protect the property left behind.

Can we *prevent* an earthquake?

Possibly. The Earth's crust is broken into various large plates, which slowly rub and scrape against each other as they move about. The joints at which plates meet (faults) are ragged and binding so that the friction is enormous. The rocks on either side of those faults slide against each other a little, then stick in place while the pressure mounts.

Eventually, when the pressure is high enough, the fault gives, there is a sudden movement, and then another sticking in place.

At every move there is a quake, and the more sudden and extensive the move, the greater the quake. Naturally, if there is only a small amount of sticking, and therefore frequent slides, there are many small quakes, none of which does any particular damage. On the other hand, if there is a great deal of sticking so that the pressure builds up for years or even decades, there is finally one large slide that creates a giant quake, which destroys everything for miles around.

Can we reduce the friction between the plates and make movement easier?

Suppose deep wells are drilled along a fault and water forced into them. The water might find its way between the masses of rocks and make them just a little more slippery, encouraging movement and a series of harmless small quakes. Then the giant killer quake will never come.

If this, or something like this, can be made to work, human beings could set about lubricating the Earth and never again have to live in fear of earthquakes.

42 THE CHANGING DAY

PRESIDENTIAL elections come in leap years. Thus, 1980 was a leap year, as were 1976, 1972, and 1968. Presidential elections come every four years and so do leap years, and that's that.

Yet Thomas Jefferson was elected in 1800 and William

McKinley was elected in 1900, and neither year was a leap year. Leap years do come every four years, but in the course of each 400-year period, we miss a leap year three times and have to wait *eight* years for one. The years that end in 00 and aren't divisible by four hundred are *not* leap years. The year 2000 will be a leap year, but 2100, 2200, and 2300 will *not*, any more than 1900 or 1800 were.

Why is this? Well, if the year were exactly 365.25 days long, then we would add a day to the 365 every four years to keep the calendar exactly even with the Sun. We would have one leap year every four years, or 100 leap years every 400 years.

However, the actual length of the year is 365.2422 days, which is just about 365 and 97/400 days, so we need 97, not 100, leap years every 400 years. We have to skip three of the one-every-four in each 400-year interval, which makes for an additional complication to the already complicated calendar.

Is there any way to stop this complication? One thing we can do is simply wait.

The day does not stay the same length. The Moon sets up tides on the Earth. There is a bulge in the water on opposite sides of the Earth, and as our planet turns, the various land surfaces move through each. The water works its way up every shore, then down, twice a day, and each time this happens there is friction of water against land, acting like a brake on the Earth's rotation and gradually slowing it down.

The Moon's tidal influence also causes the solid rock of Earth to bulge a few inches upward on opposite sides of the Earth. The rock pushes up, then down, twice each day, and as layers of rock rub and slide against other layers of rock, that too creates friction and slows the Earth's rotation.

The Earth rotated more quickly in the past. There was

a time when the day was only 23 hours, 59 minutes, and 58 seconds long. With the day 2 seconds shorter than it is now, there were exactly 365.25 of those slightly shorter days in the year, and we then could have a leap year every four years without any interruption.

The Earth will rotate more slowly in the future. There will come a time when the day will be 24 hours, 10 seconds long. With those extra 10 seconds per day, the year will be exactly 365.2 days long, and we will need a leap year every five years without exception.

There will even come a time when the day will be 24 hours, 57.3 seconds long, and there will be exactly 365 days in the year, and we won't need any leap years at all (at least until the day lengthens still more and we have to *subtract* a day from the 365 every once in a while).

When in the past and when in the future are we talking about?

As it happens, the Earth's turning has a great deal of energy to it, and the brake applied by tidal action is very weak in comparison. The present rate of the tidal action in braking the Earth succeeds in lengthening the day by only 1 second after 62,500 years.

To put it another way, each day is a rather unnoticeable one twenty-three millionth of a second longer than the day before. (There are greater changes than that from day to day for other reasons, but these other changes swing back and forth while the one twenty-three millionth of a second per day is a steady change toward lengthening days.)

This means that it was about 125,000 years ago that we had leap year every four years without exception, when Neanderthals were the highest form of humanity. And it won't be till 625,000 years in the future that we'll need leap years every five years without exception; and about 3,580,000 years in the future before we will be able to do away with leap years altogether — for a while.

So the present system doesn't require a rapid adjustment. As the day grows longer we will, every once in a long while, have to drop a fourth leap year in the 400-year interval. With time, it will have to be done at decreasing intervals. Finally, in about 38,000 years, we'll have to drop *four* leap years in *every* 400-year interval, or one every century. In other words, *every* 00 year and not just three out of four will *not* have to be a leap year.

There's no need worrying about it, though. Astronomers will keep track, you may be sure — if civilization lasts long enough.

43 HOLD THAT ICE!

THERE ARE some who fear the possibility of a coming new Ice Age as a result of the increasing dust in the air, since that dust reflects sunlight back into space, thus cooling our planet.

But suppose we clean up the air and reduce the dust to pre-industrial levels. In that case, we leave the carbon dioxide in the air at its present slightly higher level over the past, thanks to the vast burning of coal and oil that has taken place in this century.

Carbon dioxide acts as a heat blanket, and if its concentration in the air is increased slightly, the Earth becomes just a little warmer than it was.

All told, humanity has added some 300 billion tons of carbon dioxide to the atmosphere in the past century. Not all has remained there, but enough has to increase the carbon dioxide concentration in the air by 13 percent. If all the coal and oil in the ground were burned, it has been estimated that the amount of carbon dioxide in the air would quadruple.

This doesn't threaten our lives directly, of course. Carbon dioxide is a suffocating gas, but even with all the coal and oil burned, the atmosphere would be only about 0.15 percent carbon dioxide, which would not inconvenience us.

But what about the effect of the added carbon dioxide on Earth's temperature? The amount we have put into the atmosphere since 1900 is enough to raise Earth's average temperature a little over 1 Fahrenheit degree, if it weren't for the reflecting dust. If all the coal and oil in the ground were burned, the added carbon dioxide in the air would drive the average temperature upward some 12 Fahrenheit degrees.

This is not enough to bake us to death, but what about the effect on the planet's store of ice?

Right now there are some eight million cubic miles of ice resting on various polar land regions: 90 percent of it in Antarctica, 8 percent in Greenland, and 2 percent elsewhere.

This ice melts somewhat during summer at a particular pole while new ice forms during winter. On the whole there is a balance, and Earth's load of ice has remained pretty constant over the last seven thousand years or so.

If the Earth's average temperature rises, however, more ice melts in summer than is formed in winter, and the load of ice begins to shrink. Since ice reflects more sunlight than bare ground does, the shrinking ice reflects less, so that more sunlight is absorbed by the Earth, and the temperature goes up still more. At some point, the cycle becomes vicious, and the ice begins to melt faster and faster.

In the end, if Earth's temperature rises high enough (and not very many degrees of rise may be required) Earth's entire ice load might vanish.

Is this bad? Might not barren regions such as northern Canada, Siberia, Alaska, even Antarctica, become pleasant lands, offering new regions for farming and cities? Might we not actually welcome the disappearance of the ice?

We might — until we consider what happens to the melted ice.

The eight million cubic miles of ice will, as it melts, soak into the ground and run off into the ocean. The ocean basins are not only full of water, they are actually over-flowing. Nearly ten million square miles of continental area around the present shores (the continental shelves), an area greater than that of all North America, are covered by the overflowing ocean.

If the eight million cubic miles of ice are added to the overflowing volume of water in the ocean, the sea level will inevitably rise further and cover additional low-lying areas at the rims of the present continents.

When all the ice melts, the sea level will end up about 200 feet higher than it is now. Much of the lowland areas of the continents will then be under an ocean whose level would reach to the twentieth story of New York's skyscrapers.

There have been advances and retreats of the ice a number of times in prehistoric humans' history and, consequently, falls and rises in sea level. None has been catastrophic. Humankind survived.

Now, however, it is different. Because of human interference with the environment, the change in sea level is likely to take place much more quickly than it did in the past. Furthermore, the lowland areas of Earth, which face possible drowning, contain not the few hunting bands of prehistoric days, but many hundreds of millions of people who cannot easily be shifted. The areas also contain mines and farms and cities that cannot be shifted at all.

The people of Earth cannot afford either an Ice Age or

an ice melt. In the twenty-first century, therefore, one important project facing the world's technologists will be the development of methods for constantly monitoring and delicately controlling both the carbon dioxide and dust of the atmosphere to keep the average temperature of the planet within narrow limits.

We will need to develop an air-conditioned planet — one that, into the indefinite future, need no longer fear the grinding of the glaciers nor the drowning of the lowlands.

44 THE WHITE BACKGROUND

In Essay 64, I will discuss the importance of certain rare meteorites called carbonaceous chondrites and the help they might give us in determining the manner in which life originated. In that essay I suggest that the best way to study such objects is to find them in space *before* they have burned and exploded in our atmosphere, reaching the ground only in fragments, if at all.

It may be a while, of course, before our space program allows us to engage in such a search. Is there something second best that we can tackle meanwhile?

Let's see!

Suppose that a meteorite, any meteorite, survives the passage through the atmosphere and that at least a portion of it lands on the surface of the Earth and doesn't dig itself into oblivion. Let us suppose that it is lying in plain view but that no one has seen it fall.

Can it be recognized as a meteorite if anyone finds it a year later, or ten years later, or a thousand years later?

Yes, if it is an iron meteorite, for iron doesn't occur in metallic form on Earth unless it is either a meteorite or a human artifact. We can suppose that human beings can tell whether an iron object was once a frying pan or a crowbar. Therefore, any piece of iron that isn't recognizable as of human manufacture is sure to be a meteorite.

For further identification, iron meteorites are 10 percent nickel and 1 or 2 percent cobalt, and iron of human origin is usually quite different from that in composition.

Iron meteorites are only 10 percent of the whole, however. The other 90 percent are stony meteorites (of which a couple of percentage points are carbonaceous chondrites). The stony meteorites are not much different from the stones that make up so prominent a part of the Earth's crust, and unless they are actually seen to fall, or unless they land on a part of Earth's land surface that happens to be rock-free, there isn't much hope of recognizing one.

Then too, once a stony meteorite has lain a longish time on Earth's surface, it has lost some of its value even if it is recognized. It is eroded by wind and rain and is contaminated by life forms and their products. Attempts to study them with reference to any content they may have of compounds related to life are particularly hopeless.

If only it were possible for meteorites to land somewhere on Earth where they could be recognizable at any time after they have landed, even a great many years after they have landed. And if only they were but slightly affected, if at all, by erosion and life contamination.

The odd thing is that it *is* possible.

There is a sheet of ice, anywhere up to three miles thick, that covers an area of 5,500,000 square miles, or one and a half times the area of the United States. It covers the south polar continent of Antarctica, and it has been there for 20 million years. In all that time, it has been virtually untouched by life and has not encountered liquid water — the two chief factors in erosion.

Moreover, it is *only* ice, a pure white background against which anything that is not white and not ice has either been brought by human beings or has fallen from the sky.

The Antarctica ice sheet covers about 3 percent of the Earth's surface, and since it seems reasonable that any region on Earth is as likely to be struck by meteorites as any other, it follows that about 3 percent of all the meteorites that have struck Earth in the last 20 million years have struck the Antarctica ice sheet.

All those meteorites are still there!

They are easily recognizable, whatever their chemical composition, for if they are not ice and not of human manufacture, then they are meteorites. What's more, they have not been contaminated with life no matter how old they are, and they have suffered virtually no erosion no matter how long they have lain there.

Those meteorites that fell not too long ago are lying on the surface of the ice sheet, and some of them have been found!

There is no question, then, but that the Antarctica ice sheet is the richest trove of meteorites that can be found anywhere on Earth. There are probably more *recognizable* and *useful* meteorites in Antarctica than in all the rest of the world combined.

It would be a most worthwhile project to undertake a wide search of the ice sheet for any meteorites lying on the surface, and to make use of radar or ultrasonic waves to detect meteorites that may be buried under the ice.

It may be that, for a while at least, the clearest route to knowledge about the early days of the solar system will lie not in space at all, but, like the bluebird of happiness, will be found in our own backyard.

45 NEARLY WIPED OUT

EVERY ONCE in a while people speculate about the possibility of some cosmic catastrophe destroying life on Earth. In fact, the movies are full of such things these days. But is such a thing really likely to happen? After all, life has existed on Earth for over 3 billion years, and nothing has happened to it so far. Why shouldn't this peaceful existence continue?

But the existence has not been entirely peaceful. On several occasions, there seem to have been wholesale extinctions of species. The most recent occasion, and therefore the best preserved in the fossil record, took place about 70 million years ago.

At that time, the large reptiles that had ruled the Earth for 150 million years died out completely. The dinosaurs, ichthyosaurs, plesiosaurs, and pterosaurs vanished. Other, less glamorous, animals also vanished. In fact, as many as 75 percent of all the animal species then living may have become extinct over a comparatively short period.

As for the 25 percent of animal species that survived, they may have done so by only the narrowest of margins. Many but not all of them may have died, and because they were small and fecund, they could quickly reproduce and avoid extinction. It isn't difficult to imagine that at the height of the "Great Dying" some 90 percent of all individual animals may have died.

But why?

Scientists have striven to find an answer, and there have been all sorts of theories. Perhaps the shallow seas drained away as the continents uplifted, killing off many species that depended on the ecological balance of such seas. Perhaps a certain kind of plant life died off, setting off a domino effect of animal extinctions. Perhaps the climate changed sharply for some reason. Perhaps a nearby supernova drenched the Earth with cosmic rays at a time when the planet's magnetic shield was weak or absent.

Can we ever know?

Walter Alvarez of the University of California wanted to study the rate at which mud settled out of ancient seas and turned into sedimentary rock. That might help make sense out of the primordial history of the Earth. He worked with some ancient sedimentary rocks in Italy, and his plan was to bombard thin slices with neutrons. This would change the atoms present, making them radioactive. From the nature of the radioactivity, Alvarez and his co-workers hoped to deduce something that would yield information on the rate of formation of the rock.

To test the delicacy of the method, Alvarez concentrated on iridium, a rare metal similar to platinum in its properties and present in tiny traces of less than one part in ten billion. To his astonishment Alvarez found one narrow region in the rocks where the iridium concentration was about twenty-five times as great as it was in the parts of the rock that were a little older or a little younger.

And when did this happen? Alvarez worked out the age of the thin layer of rock in which the iridium was abnormally high; that age turned out to be 70 million years. The iridium jumped in concentration just as the dinosaurs were dying! What's more, such observations have been confirmed by other people and in other places.

What does this mean? Is it just coincidence?

Well, the Earth, when it formed, picked up a certain quantity of iridium. Most of this iridium, however, settled downward with the liquid iron core at the center of the planet. The amount of iridium that remained in the Earth's crust was only a small fraction of the whole, so that the Earth's crust is poorer in iridium than the material in the universe generally is.

Might we not say, then, that the reason for the jump in iridium is that material from outer space, comparatively rich in iridium, briefly showered down upon the Earth, and in so doing killed off nearly all the life on Earth?

Suppose, for instance, that the Sun underwent a brief explosion, minor on its own scale, and that afterward it settled down to its old life. The wave of heat that accompanied the explosion might have lasted long enough to kill off nearly all of life on Earth — the larger animals being more vulnerable because they were fewer in number and could less easily find protection.

Then, when some of the material hurled out of the Sun in the explosion reached the Earth, a bit of it would settle down to the ocean bottom with its telltale richness of iridium.

The possibility that the Sun once acted up is frightening, but if it did, it might do so again at any time. However, those currently considering the problem do not favor this solution.

There seems to be a strengthening view that a small asteroid, several miles wide, collided with the Earth 70 million years ago, adding a thin film of iridium to the Earth's surface.

The collision would also have kicked up scores of cubic miles of gravel, dirt, and dust, so filling the stratosphere that for perhaps three years very little sunlight could reach the surface. Most of the plant life would have died off, and most of the animal life with it. Indeed, the Earth was almost sterilized.

To be sure, that might happen again, but we know of no asteroids on quite such a collision course, so we don't really expect a repeat in the near future. Yet asteroid orbits are continually being modified by the pull at the planets, and eventually some orbit may be modified to the point where a collision will become possible. In which case, we may have to blow it up.

46 WHITHER WEATHER

IF THERE IS one thing that is of continuous interest to us, it is the weather. We can't help that. Temperature and precipitation, storm and calm, flood and drought all affect us immediately and in the long term.

It would be nice to be able to control the weather and prevent destructive extremes but, failing that, it would be useful at least to predict the weather accurately. Foreseen damage is bound to be minimized.

Yet predictions remain fuzzy and uncertain even in the short run, and just about hopeless in the long run. Despite the fact that we now have weather satellites in orbit that send down photographs of Earth's cloud cover and storm movements from hour to hour, we are not in a position to predict accurately and with confidence.

The difficulty is that Earth's atmosphere is an extraordinarily complex system. At any given moment, the atmosphere is unevenly heated by the Sun. The pattern of uneven heating changes continually as the Earth rotates on its axis every twenty-four hours, alternating day and night, and as the Earth revolves about the Sun, shifting the seasons.

Then we have a wide-spreading ocean that evaporates in sunlight and produces clouds, which form and move irregularly, complicated by the existence of irregular land areas where evaporation is much less than over the sea.

All these unruly changes in temperature from place to place in ever-shifting patterns produce winds in speeds and directions that vary in endlessly complex fashion.

One thing that might help our understanding would be to simplify the atmospheric system somehow — to reduce the unevenness of the heating, or slow the planetary rotation, or straighten the axis, or remove the cloud cover, or eliminate the irregular land masses, or empty the oceans. We could then study a more regular pattern of air movements and temperature changes, come to understand it well, add the complicating factors one at a time, note the modifications and complexities each introduces, and thus, eventually, reach a real understanding of our weather.

But how can we possibly simplify the Earth?

By seeking out *other* planets as atmospheric laboratories.

Rocket probes have already studied four distant planets with atmospheres: Mars, Venus, Jupiter, and Saturn.

Mars rotates in a little over twenty-four hours but, because it is a smaller body, its surface moves only a little over half as quickly as Earth's surface does. It is unevenly heated, as Earth is, and its axial tilt is much like Earth's, but because it is farther from the Sun its year is twice as long and its seasonal changes are slower. Because it has very little surface water, it has almost no cloud cover. Mars's very thin atmosphere should therefore have a weather pattern simpler than Earth's.

Next consider Venus. It is close to the Sun, but its cloud cover and the nature of its thick atmosphere give it a surface temperature that is fairly even over the entire planet.

It has no surface water at all, and neither temperature nor cloud cover varies significantly with place or time. What's more, the planet rotates very slowly so that it does not significantly experience the Coriolis effect that causes Earth's circular storms: its hurricanes, typhoons, and tornadoes.

As for Jupiter, it also has a permanent cloud cover and an even surface temperature overall. Jupiter, however, unlike Venus, rotates very rapidly, turning its vast bulk in only ten hours. Underneath Jupiter's thick atmosphere there is, apparently, a surface of liquid hydrogen, a kind of unbroken planetary ocean.

Saturn is like Jupiter, but less extreme. It is smaller, rotates more slowly, is colder. It represents a world with all the features of Jupiter, but somewhat simplified.

So far, we have gathered only bits and pieces of the patterns of atmospheric change on these four planets. Eventually, though, if the program of space exploration continues, we should know a great deal about each. With added knowledge we can compare what happens in the case of fast, medium, and slow rotations; of thick and thin atmospheres; of entirely solid and entirely liquid surfaces; of uneven and even heating. We will learn more about the general laws governing atmospheric movements from studying all four planets than we possibly could from studying any one planet alone.

Then we can apply that knowledge to our own atmospheric system, which is probably more complicated than that of any of those others — and understand it better, too, so that we can become expert weathermen on Earth.

It may seem odd that we must visit other worlds to understand our own — but it is a common saying that travel is broadening.

47 STRAIGHT THROUGH

How DO WE communicate with people beyond the horizon? What can be made to follow the curve of Earth's surface?

Of course, we can send electrical signals through wires around any curves. In the nineteenth century, the world strung copper wires across the continents and ocean floors and was united through telegraphy. But that takes a lot of copper and a lot of maintenance.

We could send lightwave signals and do away with wires, but lightwaves move in a straight line and won't curve around the Earth's bulge. We would have to set up relay stations, or place mirrors in orbit to make that work.

Radio waves, like lightwaves but a million times longer, do better. They travel in straight lines too, but the upper atmosphere contains regions rich in charged particles (the ionosphere) that tend to reflect the radio waves. It is as though there were natural mirrors in the sky. That makes it possible to send radio signals long distances, and in the twentieth century the world was united without wires.

However, the ionosphere is affected by the solar wind. When the Sun produces flares, an electrical storm that disrupts radio communication can take place.

Better yet, short radio waves (microwaves) can go right through the ionosphere, and be amplified and sent on by communications satellites. As communications satellites improve, signals will be sent from place to place on Earth with so little trouble that it would seem unreasonable to ask for anything better.

But what about signals that go straight *through* the body

of the Earth, and don't have to be relayed or reflected in any fashion?

What can go through the Earth itself, however? Light certainly can't. Radio waves can't. We can't even string wires through the Earth to carry electrical signals.

One thing that does travel through the body of the Earth is an earthquake wave, but it takes a very hard blow to set the Earth to vibrating perceptibly. It doesn't seem sensible to imagine ourselves slamming the Earth in Morse-code rhythms. Besides, earthquake waves follow curved paths in going through the body of the Earth, and that can be very confusing.

On the other hand, certain massless subatomic particles called neutrinos travel at the speed of light and go through matter as though it weren't there. A beam of neutrinos could travel through trillions of miles of solid lead and come out the other end just about unaffected. Neutrinos come from every direction, and almost every neutrino that reaches us passes right through the Earth in less than a twentieth of a second (and through us if we are in their path).

This doesn't mean that neutrinos can't be detected at all. One out of many trillion neutrinos may occasionally combine with an atomic nucleus and induce a detectable change.

Thus, huge vats of cleaning fluid, made up of molecules that include chlorine atoms, can serve as a "neutrino telescope." Occasional neutrinos interact with the chlorine atoms and produce radioactive atoms of the gas argon. The argon atoms can be swept out of the fluid and the radioactivity measured. Such neutrino telescopes can be placed in mines, a couple of miles under the Earth's crust. In that case, nothing can reach them *but* neutrinos, and, in this way, neutrino-producing reactions deep in the Sun's core can be studied.

Scientists can produce neutrino beams without much

trouble. Someday it might be possible to send them out in Morse code or in more complicated modulation. The day may come when improved neutrino telescopes, using water rather than cleaning fluid, will be placed all over the Earth. Eventually television sets might be built that would incorporate the equivalent of neutrino telescopes and convert the signals directly into sight and sound.

If this could be done, communications satellites would be unnecessary and so would relay stations of any sort. Any two points on Earth's surface (or in mines, or under the sea) would be connected by a mathematically straight line along which neutrinos move at the speed of light. There is no way to communicate more quickly.

For that matter, neutrinos move in a straight line throughout the universe (except insofar as they follow the curvature of space itself). They are unaffected by the electromagnetic fields and dust clouds that can disrupt or block microwaves and light.

In the end, then, it may be through neutrino beams that communications among worlds would be carried out.

Perhaps that is why we aren't detecting signals from other intelligent civilizations out there. We're looking for beams of microwaves, and perhaps we should be looking for beams of neutrinos.

48 PLAGUE FROM OUTSIDE

VAST CLOUDS of dust and gas exist here and there in interstellar space. In the last ten years, astronomers have discovered, rather to their surprise, that the gaseous part does not consist entirely of single atoms.

Somehow, although the atoms are spread far apart by earthly standards, they do manage to get together and combine. Molecules made up of as many as seven atoms in combination have been detected.

Most of these molecules are built around the carbon atom, as are the molecules in living tissue. Actually, this is not surprising. The carbon atom is the most versatile of all atoms, building up intricate chains and rings of itself to which a variety of different atoms can be attached. Atoms in space have the properties of atoms in the body and can be relied on to follow the same rules.

But what is the outcome in those gas clouds? Would the molecules build up into greater and greater complication? Might there not be some molecules present in such small quantities as to be undetectable, but that have grown so complicated as to be alive?

In short, are there germs out there in space that might some day threaten us?

It seems the very stuff of science fiction, but Fred Hoyle (a first-class but rather unorthodox English astronomer) has considered it seriously. Never mind the clouds in interstellar space. What about the comets within our own solar system?

Comets are rich in carbon, hydrogen, oxygen, and nitrogen — the four elements that make up 99 percent of living tissue generally. Might they not combine in the comet as they combine in gas clouds, and do so even more rapidly in the comet where they are in closer contact to begin with? Hoyle suggests that molecules might build up in comets to the point of developing into living microorganisms.

If so, these microorganisms would be spread through space when the comet passes near the Sun. Some of the comet would then be heated into dust and vapor and the solar wind would sweep that dust and vapor into a huge tail, which is a spectacular feature of large comets near the Sun.

The Earth passes through comet tails now and then; it passed through the tail of Halley's comet in 1910. The tails are too thin to affect us in any way — unless we possibly pick up a few germs from them — new germs to which the human body has not developed resistance.

The 1918 epidemic of Spanish influenza killed 30 million human beings. Could it have come from Halley's comet, do you suppose? And the other plagues that have suddenly swooped down on humanity. Are they comet-born? What about the Black Death? One can't help but think of Michael Crichton's best seller, *The Andromeda Strain*, which dealt with a deadly plague picked up in space.

Yet despite Michael Crichton — and despite even Fred Hoyle — I don't think it can happen.

People tend to underestimate the manner in which evolutionary processes produce a match between an organism and its environment. You don't think a human being could get along on just any planet he or she happens to stumble on, do you? The planet has to be just the right size for a comfortable gravity; at just the right distance from the right kind of star to have a comfortable temperature; with just the right kind of atmosphere; with water containing nothing undesirable; with a food supply that isn't poisonous or foul-tasting and is digestible and nourishing.

There's not one object in the solar system suitable for us — except Earth — and surely not one object in a billion in the universe as a whole that we would find endurable, let alone comfortable.

And so it is with germs. Parasites evolve into perfect fits with their hosts. They invade a particular host and live on it in a combination as close as a key and a lock. A parasite is generally confined to one organism or one small group of organisms and cannot possibly live in others, just as we cannot live on Jupiter. Do you suppose a human being can catch the Dutch elm disease, or that an oak tree can? For that matter, do you suppose that either an elm tree or an oak tree can catch our colds?

Even if it were possible for microorganisms to develop on a comet, what are the chances that they would have developed the precise structure to make it possible to invade human cells and produce the symptoms of flu? You have more chance of throwing a million successive snake-eyes, using honest dice.

Well, then, where did the Black Death and the Spanish influenza come from? From mutations of preexisting germs that are thoroughly familiar.

So cross off one possible horror that may be keeping you awake nights.

---

49 LISTENING TO THE STARS

EVER SINCE 1960, astronomers have occasionally scanned the sky for signals of intelligent origin. They looked for bursts of radiation of some sort — microwaves, for instance — which were neither perfectly regular nor perfectly random. They found none.

Now NASA proposes to spend five years and twenty million dollars to listen to every Sunlike star within a few hundred light-years of ourselves.

Is there really a chance astronomers may hear something?

Yes, though not a big one.

Consider probabilities of a particular star being much like the Sun and having a planet much like the Earth, on which first life, then civilization might develop. It is possible to argue that there may be as many as 50 million civilizations, each more advanced than ours, scattered through the Milky Way Galaxy of which our Sun is part.

That sounds like a great many, but it means that only about 1 out of every 3000 stars has in its system a planet that bears a civilization.

On the other hand, this would be true only if the civilizations were long-lived — if once one of them got started, it continued to exist and develop indefinitely. Judging by the history of our own civilization, however, it seems to be an unlikely event. It may be that once a civilization reaches the nuclear-weapon level, it may wipe itself out by using those weapons or through other catastrophes made possible by technological advance, such as overpopulation and chemical pollution.

If one were completely pessimistic, one might decide that there were no advanced civilizations at all because all committed suicide, as ours may be about to do.

But surely some would escape. Even if only one civilization out of a million managed to overcome the tendency toward hostility and violence, and developed a cooperative mode of life that enabled it to move forward indefinitely, that would still mean there were fifty advanced and long-lived civilizations in existence right now.

But there is a difference between fifty civilizations and 50 million. If fifty civilizations are scattered evenly through the Galaxy, there is only an even chance that one of them might be as close to us as 30,000 light-years. If there were 50 million, there is an even chance that one is as close to us as 300 light-years. To reach us, a signal from 30,000 light-years away would have to be ten thousand times as intense as one from 300 light-years.

The fewer civilizations there are, then, the less likely it will be that we will detect anything.

But wait — that assumes that a civilization established in

a particular planetary system will remain isolated there for all its existence. Is that assumption valid?

We ourselves are stuck in our solar system because we have not developed any practical method for interstellar travel. Yet might we not, given another few centuries of steady advance (if we avoid civilization-suicide) develop the technique?

And if we could do it, might we not reasonably assume that any civilization more technologically advanced than ours would have interstellar travel and be able to colonize any habitable worlds it encounters?

It would be natural to suppose that those civilizations which survive to the point of developing interstellar travel, however many they may be, will all have done so only because they developed cooperation and nonviolence to a high pitch. Presumably, they would not attempt to occupy those worlds that, like ours, were already occupied by intelligent species. What's more, if, in their colonization ventures, two civilizations encounter each other, might they not peacefully form a "Galactic League of Civilizations"?

In that case, however few the existing civilizations, millions of habitable worlds may now be organized by them.

If astronomers, listening to the stars, detect nothing, it could mean that there are no advanced civilizations out there — or, if there are advanced civilizations, none is near enough to detect because interstellar travel is, after all, an insoluble problem — or, if there are millions of civilized planets and even a galactic league, none happens to be sending out signals, either deliberately or inadvertently, that we can pick up.

If, by some chance, we *do* pick up signals of intelligent origin, then we would at least know that it is possible to survive the perils of technology and that we may be able to do it ourselves. We would also become aware of the possible existence of a galactic league in which, someday, we may find ourselves.

Finally, it would give us a target toward which to send our own signals. I even know what we should say, if we could figure out how.

"Help!"

NOTE: This is the only essay in the book that did not appear in *American Way*. Through an unfortunate coincidence, the magazine had commissioned a major article on the very subject of communication with extraterrestrial intelligences in the same issue for which I had aimed this essay. I had to substitute something else quickly.

---

50 EYE IN A VACUUM

THE TELESCOPE was invented in 1608, and Galileo constructed one and turned it on the heavens in 1609. His primitive telescopes used poor lenses that were only an inch and a half or so across, but he made wonderful discoveries with them, for no one else had ever looked at magnified heavenly objects.

We've made tremendous strides in the 372 years since. At Mount Palomar there's a telescope in which great quantities of light are gathered and concentrated through the use of a virtually perfect parabolic mirror 200 inches wide. In the Soviet Union there's a telescope with a mirror 236 inches wide.

Those are probably the last of the dinosaurs. In Arizona, something new has begun operation — not one large mirror but six smaller ones, each 72 inches wide, all working in unison. A number of small ones do the work of one much larger one — small mirrors are easier to grind perfectly, are much less massive and therefore less distorted by gravitational pull, and can be adapted to work much more flexibly. Of course, it would take a computer to organize them to work together, but we now have computers that can do it.

All the telescopes, from first to last, however, labor under the same enormous disadvantage. They are all here, on the surface of the Earth. Above them is an ocean of air that is sometimes misty and foggy, sometimes outright cloudy. Even perfectly clear air absorbs about one third of the starlight and blocks most of the wavelengths outside visible light altogether.

Then too, the air trembles with temperature change, and that blurs the stars and planets constantly. Man-made lights in cities and on highways create a growing glare that is harder and harder to avoid, while man-made dust and smoke make the atmosphere dirtier and harder to see through. Even telescopes on mountain tops and on desert heights don't escape entirely.

The only answer is to get out from under the atmosphere. All sorts of astronomical observations have already been made, first from rockets probing above the atmosphere and then from satellites that stay above for months and years. On Skylab, astronauts actually used small telescopes to make observations.

Since 1962 people have dreamed of putting a *large* telescope into orbit, one with a mirror as large as 120 inches wide.

A telescope in orbit about the Earth would be in free fall and wouldn't feel the pull of gravity. It would need to weigh only a twelfth as much as a similar instrument on the surface of the Earth, since it wouldn't have to be propped up against gravitational distortion. Without the pull of gravity to fight against, it would respond much more readily to automated control and would have to be visited only occasionally for adjustments, repairs, and supplies.

Without atmospheric interference it would see things more clearly than a similar telescope on Earth's surface would. It could, in fact, make out objects a hundredth as bright as the largest Earth-based telescope now in existence, so that records could be taken ten times as far out into the universe as is now possible.

Then too, the entire spectrum could be seen — *all* the waves of radiation, not just those that get through our atmosphere.

What will we see with the "Large Space Telescope"? It's hard to predict. After all, Galileo couldn't have predicted that he was going to see mountains on the Moon, satellites of Jupiter, phases of Venus, spots on the Sun, and stars in the Milky Way.

We should be able to see the inner regions of giant star clusters and of galactic centers more clearly with a telescope in space. Perhaps we'll find out whether these objects really have black holes in them. That is important, for black holes may tell us if, when, and how the universe will end.

The Large Space Telescope will be able to photograph ten times as many distant galaxies as it is now possible to photograph. It may give us the necessary information to understand in detail what is happening up to ten billion light-years away and, therefore, ten billion years into the past. It might teach us what we need to know to understand how the universe began — whether it will expand forever and pass away — or whether it will someday contract again and begin all over.

It may give us the information we need to work out new and more fundamental laws of nature, new angles to old forms of energy, or new forms of energy altogether.

We don't know. We haven't the faintest idea what might come. That's what makes the prospect so exciting.

51 THE WATCH IN SPACE

ON AUGUST 10, 1972, a meteor streaked through the upper atmosphere above western United States and Canada — through and out again. It was just a near-graze; no harm done; but it was close enough to be disconcerting.

It was moving at a speed of just over 6 miles per second relative to the ground when it entered the atmosphere over Utah and had speeded up to 10 miles per second by the time it left the atmosphere somewhere over Alberta. At its closest approach to Earth's surface, in southern Montana, it was 36 miles high. The estimate is that it was 43 feet wide and weighed 4000 tons.

Had it come a little closer it might have curved downward and struck with considerable force, becoming a meteorite, which is what meteors are called after they strike. It would have made a pretty big hole in the ground.

Meteors that size, and larger, have struck the Earth in the past. About 19 miles west of Winslow, Arizona, is Meteor Crater, a hole about 600 feet deep and 4000 feet wide. It was gouged out by a meteorite that struck anywhere from 5000 to 50,000 years ago. That meteor was probably some twelve times larger than the one that grazed us in 1972.

What else is out there in space?

There are asteroids of varying size whose orbits can bring

them closer to Earth than any planet ever comes. The largest is Eros, which was first sighted in 1898. It is 15 miles wide, but it never comes closer than 14 million miles to Earth. It's safe.

Since 1898, however, a number of other asteroids have been discovered that are smaller than Eros, but also come considerably closer to Earth, if both bodies are located properly in their orbits.

In 1937 a large piece of rock, given the name Hermes, was sighted at a distance of a mere 400,000 miles from Earth. It is 1 mile wide and may be nearly 2 million times as massive as the rock that grazed us. If its calculated orbit is correct, it could come to within 200,000 miles of us — less than the distance from the Moon. Where is it now? No one knows. It was never seen again, so all we can say is it's out there somewhere.

There may be thousands of rocks smaller than Hermes, but still large enough to cause considerable trouble. They're all out there somewhere, and any one of them might strike some day. Some have done so in past history, gouging craters that wind and weather and living things have worn away — or splashing into the sea and stirring up a tidal wave.

A few may have struck since humankind has appeared on the Earth, as did the one that gouged out Meteor Crater. Till recently, though, the human population has been sparse and its works trifling. Even a large gouge or splash might not have done much harm. (However, see the end of Essay 45.)

Now, however, the Earth is networked with cities and with the gigantic works of man. A meteorite strike almost anywhere would gouge out a crater, or thrust up a wall of water against some coast, that could not help but cost an incredible amount in life and property.

It may be, then, that one of the important occupations

of the future will be that of manning the "meteor watch." A Moon-based telescope, a manned space station, unmanned satellites (or all three) might serve to scan the space around Earth perpetually for any sign of anything of appreciable size.

Anything sighted would be tracked as far as possible and its orbit worked out. These meteoroids (which is what they are called in space) would be catalogued, with their orbital characteristics carefully listed.

Every once in a while, some meteoroid in the catalogue will make another close approach to earth. Such a return will be expected and the various telescopes, or other detecting instruments, will be on the watch to make sure the meteoroid is safely on course, or to see whether some unexpected or miscalculated effect has altered it slightly — and dangerously.

There is bound to come a time when the figures will indicate something close to collision course. We can hope that by that time the art of space gunnery with hydrogen missiles (or something more sophisticated) will have advanced to the point where the meteoroid can be intercepted in short order.

Even if it is shattered and vaporized into a trillion pieces, those pieces may still hit the Earth. In that case, however, each piece will burn in the air without doing any harm, and the part of the Earth under the atmosphere-entry will be treated to a magnificent display of harmless aerial fireworks.

If even a single meteoroid is shot into harmlessness and prevented from gouging out a region or drowning a coastline, the saving in property damage alone would probably pay for a hundred years of space exploration. As for the saving in life — how do you count that?

52 THE SAFETY OF DISTANCE

SOME YEARS AGO the populous regions of northeastern China suffered a severe earthquake, and damage both to lives and property was enormous. Yet where I sit, ten thousand miles away, no faintest tremor could be felt. At a distance of even five hundred miles, life would carry on normally.

The most damaging earthbound cataclysms of nature are purely local in effect. Otherwise, civilization could scarcely have developed and life itself might exist only precariously on our planet.

There are much greater cataclysms outside Earth — and, fortunately, much greater distances to act as much more effective insulation. The Sun is a gigantic ever-fusing hydrogen bomb, but because it is 93 million miles away, it is a benign agent for light and warmth.

Our atmosphere, to be sure, absorbs much solar radiation that might be deadly even at this distance. But place our planet at half its distance from the Sun, and our atmosphere would not save us. The ocean would boil and life would come to an end.

There are still greater cataclysms beyond the solar system. There are stars, and even whole galactic centers, that explode and collapse; and there are quasars that, in a tiny space a few light-years across, develop a hundred times as much energy as whole ordinary galaxies millions of times their volume.

Around such mad outpourings of energy, life as we know it would be made impossible for light-years in every direction — but we are safe, for these explosions are separated from us by distances of from hundreds of light-years to billions of light-years.

Distance is our salvation, and we seek it automatically for man-made dangers too, real or fancied. Children run from a haunted house and are no longer afraid when they have put a few hundred yards between it and themselves.

Soldiers leave the battlefield for rest and recuperation behind the lines — a movement of a few dozen miles may be enough to surround them with a world at profound peace. Victims of a repressive government can cross a boundary line and feel safe.

But what if the ancient refuge of distance is destroyed? What if nowhere on Earth is there safety? That is a new and unaccustomed horror for humankind, different from anything that has ever existed. What if there is a world government, harsh and oppressive, from which fugitives can find no haven of safety? The solution would be to prevent the establishment of a world government (and this is the only argument against one that I find difficult to answer).

What if hydrogen bombs can be used to poison the atmosphere and expose every bit of Earth's surface to deadly radiation? The solution would be to prevent hydrogen bombs from ever being used in that way.

But what if a purely negative solution is insufficient?

What if nuclear power stations intended as peacetime sources of energy possess the potential to poison the Earth generally? And what if we cannot abandon nuclear power without the cataclysm of an energy shortage that will reduce our civilization to chaos?

Or what if experiments in genetic engineering produce the possibility of a deadly worldwide epidemic from which there is no escape? And what if we ought not to abandon such experiments because they might produce enormous benefits, as I have suggested elsewhere in this book?

How do we resolve these dilemmas? Why, again, by the safety of distance! If the Earth is too small to supply the necessary distance, then we must get off the Earth, for the distances are greater there.

Suppose, for instance, that there were uranium or thorium on the Moon. Might we not be able to build nuclear fission power stations there, beam energy to satellites in synchronous revolution about the Moon, from there to other satellites in synchronous revolution about the Earth, and from there to Earth? It would be very complicated and inefficient, but think how much safer for Earth. Plutonium theft would be unlikely; indeed, radioactive waste disposal would be no problem, and any meltdowns or radioactive escape would endanger only the volunteer workers and technicians who understand the risk and are willing to chance it.

And suppose we have a space colony devoted to conducting experiments with genes, experiments that might produce a deadly disease. The people on the colony might die, a risk they would knowingly take; but the epidemic would be safely insulated. Think how much more easily a space colony could be controlled and how much less likely it would be to serve as a source of infection than a similar laboratory located, say, in Cambridge, Massachusetts.

Earth can never be made entirely safe — but it can certainly be made a little less unsafe if the worst dangers are put on the other side of a stretch of vacuum — at a distance.

53 FAREWELL TO WORLDS

THE ONLY examples of life that we know originated, scientists believe, billions of years ago in the outermost layers of the Earth. That's where the basic matter from which life could form was present in sufficient quantity and variety, and where it was properly bathed in the energy of sunlight.

Somehow that has made us feel that life and worlds are inseparable. If other forms of life are to be found out in space somewhere, we take it for granted they are to be found on the surface of other worlds. If humankind is to spread outward beyond the surface of the Earth, it is to be through the colonization of the surface of other worlds.

Yet almost all the matter in worlds is virtually useless as far as the immediate needs of life are concerned. On Earth itself we certainly need the atmosphere — all of it — for our minute-to-minute needs; we need the bodies of salt and fresh water; and the outermost several miles of the Earth's solid crust — but that's all. Everything below those several miles might as well not be there as far as any immediate use we can make of it is concerned.

To be sure, all that vast untouchable quantity of world beneath Earth's skin affects us and is even vital. It makes up almost all the mass of the Earth and therefore supplies almost all the intensity of the gravitational field. If the Earth were a hollow shell (as some cultists say it is) we would weigh only a fraction of what we weigh now; and if the shell were thin enough, the Earth would have so small a gravitational field as to lack an atmosphere, an ocean — and life.

You see, then, that a large, solid world is indeed necessary for life to begin — but once life has come to be, and has matured, and starts to reach out for regions beyond the world of its origin, need it confine itself to other worlds only?

Professor Gerard K. O'Neill of Princeton University thinks not. In a detailed consideration of the astronomical, structural, and biological aspects of the situation, he points out that it is perfectly possible (if rather expensive) to build artificial cylindrical structures in space that can house sizable populations of human beings.

Initial structures of aluminum and glass might be only about 1100 yards long and 110 yards wide, but O'Neill has calculated the requirements and costs of larger structures — up to 20 miles long and 2 miles wide.

Though the gravitational pull of such structures would be negligible, that would not matter. No one would rely on the cylinders to hold air, water, and soil to their outside surfaces by gravitational pull. The necessities of life would be on the *inside* of the cylinder, and would be kept there by virtue of being tightly enclosed.

For that matter, there would be a substitute for gravity in the form of a centifugal effect induced by the spinning of the cylinder. Small cylinders might spin at a rate of 180 times an hour and large ones at a rate of only 30 times an hour. As a result of that spin, soil and water would be pushed against every part of the inner surface of the cylinder. So would human beings, who would feel the effect as the normal pull of gravity, provided they are near the cylinder's inner surface.

The number of people who could inhabit individual colonies can vary, O'Neill estimates, from 10,000 to 20 million, depending on the size of the cylinder. The large ones can be constructed in such a way as to have mountains along the inner surfaces of the long end of the cylinder, with cloud formations too. The larger the cylinder the more nearly like home it would seem to be — except that it might be carefully started with no insect pests, no pathogenic bacteria, no sources of pollution.

Where do the materials to build these colonies come from? For a large cylinder, some 15 billion tons of material might be needed, and there would be an understandable reluctance to despoil Earth of such quantities of badly needed resources. But we don't have to turn to Earth.

As luck would have it, we have the Moon out there, only a quarter of a million miles away. It is one-eightieth the mass of the Earth and lifeless. It could be used as a virtually limitless supply of materials to build any number of colonies. The Moon material contains soil and rock that can be used as such, and can also be treated to extract metals and other elements for all possible uses.

About the only necessary elements not present on the Moon in sufficient quantity are carbon, nitrogen, and hydrogen. That would have to be contributed by Earth (which has a considerable supply) until the colonies develop to the point where they could mine other sources — comets, or some asteroids.

It may all turn out to be politically and economically unfeasible, but what a fascinating idea it is to suppose that much of the human population might someday be able to say farewell to worlds and live comfortably inside orbiting space cylinders instead.

54 CATCHING A BIG ONE

As WE have seen, it's quite possible that within a few decades humanity will have established a mining base on the Moon. The crust of the Moon can be smelted to yield oxygen, glass, ceramics, and a variety of metals. We can use such material to build all sorts of structures in space — power stations, laboratories, observatories, factories, human settlements.

Unfortunately, the Moon is low in a few light elements that, in themselves or in common combinations, are easily vaporized. The Moon's gravity is too weak to hold these vapors, and if they were ever present on the Moon, they are gone now.

The Moon is short in hydrogen, carbon, and nitrogen in particular, three elements that would be extremely useful — in fact essential — to all space endeavors. These light elements are in plentiful supply on Earth, which could contribute them by the ton to get things started.

As time goes on, though, and as space is increasingly exploited, with more and more of everything requiring more and more supplies, Earth is bound to feel the strain and become parsimonious with its light elements. In that case, if space is to continue to be an expanding human habitat, some other source of light elements must be found. But what other source?

Besides Earth, only two worlds in the inner solar system contain light elements at all — Venus and Mars. Venus is the closer of the two, but it is ferociously hot, and no base can be established on its surface. Even if one could be established, Venus's surface gravity is nearly equal to that of Earth, and lifting material from it would be an expensive proposition.

Though Mars is farther from us and colder than Venus is, it has some light elements in convenient solid form (frozen water, frozen carbon dioxide) that Venus does not, and its surface gravity is only two fifths that of Earth.

We might imagine a permanent station established on Mars, the chief purpose of which would be to launch cargoes of ice and frozen carbon dioxide into an orbit that would carry them, eventually, close enough to the Moon for them to be picked up easily.

Yet the day will come when human beings will settle Mars and its vicinity, and it would be a shame to rifle it in advance of the resources that the Mars settlers will need.

Yet where else can we turn? The outer solar system beyond the asteroid belt has vast quantities of light elements, but the worlds out there are so enormously distant that it may not be practical to use them as a source for a long time to come.

But not everything in the outer solar system remains there permanently. Very far out from the Sun is a belt of countless billions of small bodies built up almost entirely of the light elements. Most remain out there, but every once in a while, collisions, or small gravitational pulls that mount up, send one of them into an elongated orbit that brings it into the inner solar system, around the Sun, and then out again.

In the inner solar system, the heat of the Sun partly vaporizes the frozen bodies, raising dust and vapor — and we see that as a comet. Over and over again, a comet will follow its elongated orbit, losing more and more matter to solar action until, eventually, it is all gone, or until only a small rocky core is left.

Why let all those light elements go to waste?

We can picture space settlers lying in wait for such comets. Space-based telescopes would detect large comets, a few kilometers wide, after they passed Saturn's orbit, and the really big ones would be bonanzas.

The comet's orbit would be plotted, and in the long months it would take them to reach the inner solar system, the space settlers would have placed a ship at some rendezvous point in space. A landing would be made on it, and it would be outfitted with some sophisticated propulsive mechanism that would force it out of its orbit. It could perhaps be made to spiral down to the Moon and into the shadow of some inner crater lip at the lunar poles, where the light of the low-lying Sun would never penetrate.

It would be like hooking, maneuvering, and landing some giant sea monster — and catching a big one would bring enormous benefits to space society, for its cubic kilometers of light elements, carefully hoarded, might last for decades. By then, of course, some other large comet will have heaved into view.

It may be that "comet-fishing" will keep our space settlers prosperous and expanding until such time as the outer solar system can be penetrated and exploited.

55 THE BRIGHT AND HEAVENLY MILLS

In 1804, when the Industrial Revolution, at its beginning, was boring into England's economic structure, the poet, William Blake, already viewed its consequences with hatred and foreboding. In a preface to a poem entitled "Milton," he contrasted the works of God and Devil:

> *And was Jerusalem builded here*
> *Among those dark Satanic mills?*

Ever since then, the world has not lacked for those who decried industrialization and all its works; who have been pained by the mass production of undistinguished artifacts, the busy-busy of meaningless noise and waste, the sprawl of concrete and asphalt that endlessly swallows the green land, and the interminable production of effluvia (gaseous, liquid, and solid) that progressively poison the air, water, and soil that keep us alive.

Why don't we get rid of industrialization? Because we cannot. We depend on it too much.

The mass production of undistinguished artifacts enables

vast numbers to have something. They are not works of art, but the alternative is nothing. The rest of the industrial mess provides jobs, a higher standard of living, lengthening life, and increasing comfort.

The fact is that most of those who denounce industrialization are careful to keep its benefits. They are themselves usually among the better-off whose way of life depends upon the dark, Satanic mills at every point.

It is also a fact that the world has always been ready to risk the dangers of industrialization for the sake of its benefits. No sizable society has ever voluntarily given up its industry, and unindustrialized societies continually strive to develop industry.

Is there no way out of this dilemma? What about space?

Among the many benefits of space advanced by those who, like me, are enthusiastic about the expansion of humanity into the realm beyond the atmosphere are its uses in connection with industry. Space might be the site of a new industrial revolution, with large industrial complexes on the Moon, or in orbit about the Earth.

The advantages are many (if the political and economic hurdles can be overcome). A totally artificial environment geared to the necessities of the industrial process can be designed. Advantage can be taken of properties of space that can't be duplicated easily (or at all) on Earth — no atmosphere, little or zero gravity, high temperature and radiation or low temperature and radiation, and the lack of an ecological system to distort or destroy.

We can envisage the twenty-first century as a time in which Earth's industrial system will be gradually lifted into orbit. And as more and more of it is lifted beyond the atmosphere, less and less will be left on Earth's surface. What the process will give Earth will be even less important than what it will take away, for it will take away the noise, the sprawl, and the pollution.

The waste products that inevitably result from industrial processes would be more easily recycled in space-based plants designed from the outset for such a purpose. Where this is not possible, waste products discharged into space have, literally, uncounted millions of times as much room into which to dissipate as they would have on Earth's surface, and there would be no living things to disturb in the process.

The factories in space, largely automated and computerized, would require relatively small contingents of human beings to supervise and maintain production; these could be drawn largely from the space settlements that would also exist in orbit.

The space settlements, and Earth itself, would retain all the benefits of industrialization and be freed from its disadvantages, since industry *would* exist, but would exist *elsewhere*.

Earth would become an increasingly agricultural and pastoral world, with large areas devoted to park and wilderness. Humanity could then construct a stable and workable ecological balance on the planet, once the pressure of ever-increasing population and ever-increasing surface industrialization is removed.

However pleasant and Earth-simulating space settlements might be, they would never duplicate Earth's vast expanse and natural grandeur, or the complexity and variety of its ecology. A nonindustrial Earth might therefore become a great tourist attraction and retirement home.

Earth would be an enormous suburban, rural, and wild world, for which the space settlements would serve as cities, and the orbiting factories would be the bright and heavenly mills (no longer Satanic) from which the comfort and security of humankind would be drawn.

56 PIE IN THE SKY

IN ESSAY 53, I mentioned the possibility of space settlements — of large numbers of human beings making their homes in artificial worlds orbiting the Earth at distances as far as that of the Moon. In Essay 55, I pointed out that industrial plants might also be placed in orbit about the Earth.

With people and industry moving away from Earth's surface, what will be left? Other people and other industry, of course — and *farms*. Surely Earth would remain the great, green, food-spawning globe that would feed the multiplying people of space. Or would it?

Perhaps not.

The various space settlements would have their inner surfaces lined with soil built of lunar materials and carefully designed for plant growth, with adequate supplies of trace minerals — neither too little nor too much. Sunlight, controlled by great mirrors and adjustable "venetian blinds" in the windows of the space settlements, could supply light of just the proper intensity all year round. Temperature, water, and carbon dioxide would all be carefully adjusted.

In short, farming in the space settlements would not be dependent on the vagaries of the seasons of a world too large to control, but would be carefully designed for optimum conditions. The crop yield per acre in a space settlement would be much higher than on Earth, and the settlements would surely produce foods for export.

As the settlements multiplied in number, they could become the granary of humanity. Earth could afford to specialize in the cultivation of less economical foods — animal herds, for instance — so that there would be a brisk exchange of settlement grain for planetary meat.

While each space settlement might strive for balanced food production to feed its own limited population, each might also specialize in a different export product, for which its ecological balance would be carefully designed. There might even be genetic experimentation — the use of solar ultraviolet and x-rays to produce plant mutations that could introduce species with novel flavors, tastes, and other properties. From space would come plant food not only in sizable quantities, but in unprecedented varieties. We'll really have "pie in the sky," at last.

On the more theoretical side, the space settlements, if they are to make agriculture work, will have to study the science of ecological balance and manipulation with great intensity. How few species of microorganisms will satisfy the requirements for maintaining soil fertility? Which plants require what insects? Which groups of plants and animals fit together into a working whole, and which do not? The knowledge gained may well be applied to Earth's supremely large ecological structure so that Earth itself may be able to produce food in unprecedented amounts.

The side effects can be both delightful and tiresome. On the one hand, the differences between the settlements will encourage tourism, as each settlement will have its own specialized cuisine. On the other hand, incoming tourists (especially from Earth) are likely to have to undergo tedious examination and quarantine to make sure there is no inadvertent introduction of unwanted species that will upset a fragile balance.

We might go further. We can imagine space structures that are, essentially, large vessels of water. Each vessel could be illuminated intermittently by reflected sunlight from all angles, an illumination that would keep it at some appropriate temperature, while the thickness of glass enclosing the whole would be sufficient to absorb the harmful ultraviolet radiation of sunlight. Inside the vessel would be algae, one-celled plant organisms that can multiply and produce edible substances more rapidly than any other form of multicellular life.

The contents of the vessel would circulate through ancillary equipment so that carbon dioxide and various mineral nutrients would be introduced at one end, while the algae would be filtered out at the other. Periodically, there would be a major flushing, with fresh batches of algae added. The whole would be automated so that only minimal human supervision would be required.

The countless tons of algae thus collected, properly desiccated and fractionated, could produce a nourishing powder that would serve as animal feed at the very least and probably as additives for human food as well.

57 I CAN FLY; I CAN FLY; I CAN FLY

THE ANCIENT Persian monarch, Xerxes, is supposed to have offered a reward for anyone who could introduce a new pleasure to him. I haven't heard that anyone ever claimed the reward.

There are, of course, many pleasures that just don't please everyone. Mountain-climbing excites many people, and skiing excites even more, but neither one would tempt me two feet out of my way. On the other hand, some of the things that give me intense pleasure — like sitting hunched over my typewriter — may not please you.

There is one yearning, however, that may be almost universal. Surely, there can't be a child who hasn't wanted to fly like a bird and to feel an utterly controlled three-dimensional motion. Men have flown in their dreams, legends, and fantasies for as far back as our literary records go.

We *do* fly now, of course. Ever since the first balloon went up, people have been able to travel through the air. It was never, however, by way of their own muscular control. The modern airplane is so luxurious, comfortable, and safe that one might almost be in one's easy chair at home. Even if we were to make use of a small jet motor to lift an individual body into the air, one would still be *carried*.

Why can't we fly by flapping artificial wings and using our own power? Alas, our muscles simply aren't strong enough. Even light, cleverly designed aircraft like the *Gossamer Albatross*, kept aloft by human muscle power alone, require far too much effort to make flying them a pleasurable exercise. The trouble is that the air is too thin to give enough support to lift our heavy bodies against the pull of gravity. (The heaviest living organisms capable of flight weigh no more than 40 pounds or so.)

But if we colonize the Moon, there would be an important change in the situation. The gravitational pull on the surface of the Moon is only one sixth that on Earth, so that individual human beings would weigh only 20 to 30 pounds.

Eventually, when an underground lunar colony is a going concern, a large cavern may be built, in which air might even be kept at somewhat more than normal pressure. (Argon, a harmless gas that could be available in the lunar crust, might be added to the air.) The combination of low weight and denser air should make flying simple.

People could use a light harness to which tough and semirigid plastic "wings" would be added — not just to the arms but to the legs and body as well. Thin, firm supports within the plastic might be attached to the individual fingers and to each leg, after the fashion of a bat's wing.

It would be so easy to glide under such circumstances that there would be almost no chance of being hurt and, with practice, the wings could be tilted, bent, or even flapped in such a way as to make it possible to ascend and descend, to turn and bank.

Simple flying under such circumstances would be possible for anyone but, as in anything else, there would undoubtedly be those whose greater rapidity of response and superior muscular coordination would make it possible for them to carry out maneuvers beyond the reach of the average individual.

We can easily imagine obstacle races by expert aeronauts, or games equivalent to air soccer, or precision gymnastics by trained teams.

To be sure, conditions might be too cramped within the underground lunar colony for truly spectacular flights, but what of the possible space colonies built within huge cylinders miles in diameter?

There the air is not likely to be denser than Earth; rather thinner, if anything. The "gravity," however, will be the result of the centrifugal effect set up by the cylinder spin about the long axis. This "gravity" will probably be lower than the gravitational pull on Earth, and even if it isn't, it will drop off as one moves away from the inner surface of the cylinder toward the central axis. It would be zero at the central axis.

If there were mountains built up over portions of the inner surface of the space colony, neither temperature nor air density would drop as one climbed, for those would be uniform, or nearly so, throughout the colony. The "gravity" would decrease, however, and from the mountain top, if not from the valley, one could launch oneself into the air and climb — and as one climbed, find it easier still to climb farther.

In such a colony, there would be miles within which to maneuver, with an added dimension of pleasurable complexity, since the "gravity" would change perceptibly as one swooped up and hurtled down. In such a world, who would want merely to walk?

There's the new pleasure, Xerxes — but not on Earth.

58 PLAY BALL!

No one can watch a professional ball game without marveling at the expertise of the players, once one stops taking it all for granted.

Watch the outfielder take one look at the soaring ball, turn his back and run, and then, without ever breaking stride, catch that ball over his shoulder. How did he know where it was going?

How does the tennis ace know just where to run at top speed to have his racket meet the ball? How does the football receiver evade and jump just right to catch the pigskin?

We might think it is nothing less than magic were it not that we were so used to it; and yet, in a way, it is simple. The balls follow fixed rules of motion, speeding in parabolas (as modified by air resistance and wind) so that when one sees the beginning of the curve, one can estimate all the rest of it and know where the ball will be at any given time.

That doesn't mean anyone can do it. You or I would be nothing but duffers if we tried. We are, however, talking about the expert few. They may not have the faintest idea what they are doing or how they do it, but they can do it. Through long practice, through a good eye and quick reflexes, their bodies can move in such a way as to allow for the parabolic motion of the ball and for any modification by wind.

But all the body-movement games we play now are the same, in essence, as those that have been played through all of history. We have always been held to the ground, and flying objects of one sort or another have always followed the rules of inertia and gravitational pull.

Now, however, for the first time, we can foresee the possibility of a fundamental change of rules. Suppose we imagine a large sports arena in orbit about the Earth. If the arena is rotating slowly or not at all, then at all points within it there will be a sensation of zero gravity, as in Skylab.

Tennis can become a three-dimensional game! This is not to be confused with indoor handball, where the ball can bounce off walls and ceilings, but the players themselves are confined to the floor.

Imagine tennis within a large cube, with the foul lines drawn on four surfaces, and the players at the two unmarked ends opposite each other. The balls will travel, not in flat parabolic arcs, but in absolutely straight lines. And the players, perhaps equipped with small reaction motors, will not be running along the ground, but will have an additional degree of freedom.

Whereas the present-day tennis player, on the surface, must run right and left (one degree of freedom) or forward or backward (a second degree), she will, in space, have to move up or down in addition. And she must do it all in any combination.

Surely it can be done. Experienced pilots can fly airplanes in close order with Rockettelike precision while engaged in turns and loops at high speeds, so we know three degrees of freedom can be handled.

But then we can imagine ourselves moving on to something more complicated still. On the surface of the Earth, gravitational pull is always constant and doesn't change, whether the ball speeds along near the surface or is a hundred feet in the air. In free fall, gravitational pull is always constant too — always zero.

What if we can manage to change the force of the pull, however? Suppose the large spherical arena is set in rapid rotation so that, like the Earth, it will have poles and an equator. Thanks to a centrifugal effect, the inner surface, at or near the equator, could have a pull similar to that of gravity at Earth's surface. The farther one moves away from the equator toward either pole, the smaller the apparent gravitational effect, until it is zero at the poles.

If you imagine a soccer game sweeping over the inner surface of the sphere, with both goals on the equator, the players themselves, as they pursue the ball from side to side, must encounter changing gravitational pulls and must not only adjust their muscular efforts to suit, but must follow the movements of the ball in accordance with a pattern far more complicated than is required in any constant gravitational field.

In fact, in such an arena, the gravitational pull decreases as an object moves up and away from the inner surface toward the axis of rotation. A baseball flying into the air is subjected to a weakening pull. It not only rises unexpectedly high, but is twisted to the side in its flight, thanks to something called the Coriolis effect. It becomes *much* more complicated to try to judge the ball's movements.

Once space sports are developed, it isn't hard to imagine that the kind of play we now have on Earth's surface may be considered suitable only for children.

59 WHAT TIME IS IT?

THE COLONIZATION of space may introduce some unexpected changes into human society. For instance, what effect will it have on the way we keep time?

Our present system of time-keeping is a complicated mess that depends on the accidents of astronomy and on five thousand years of primitive habit. The length of the day is, of course, the period of Earth's rotation about its axis, while the year is the period of Earth's revolution about the Sun.

The two do not fit together neatly since 1 year = 365.2422 days, which complicates our calendar a great deal. Add to that the month, which originally marked the period of revolution of the Moon around the Earth, and the week, which originally marked the phases of the Moon, and things are almost hopeless.

Put it all together and we've ended with a different calendar for each year — no easy way of predicting on which day of the week some particular day of a month will fall, no easy way of calculating the number of days between two dates, and so on.

On top of that, we divide the day into 24 hours, the hour into 60 minutes, and the minute into 60 seconds. What's more, we count the hours in a day from 1 to 12 twice, giving us an A.M. and a P.M. All this is entirely because the early Babylonians, not having worked out the concept of a decimal system, used unit fractions instead and for that purpose found 12, 24 (12×2) and 60 (12×5) to work better than 10.

Once humanity moves out into space, however, what

will be the value of our complicated time-keeping system? People in space settlements will have a seasonless year, since they can adjust sunlight to suit themselves. There will then be no purpose in deliberately maintaining a year that fits the days unevenly. There will be no lunar cycle to give meaning to months and weeks. Even the day will be artificial, since the alternation of light and darkness can be adjusted at will.

If we consider people living on other natural worlds, Earth's astronomy would certainly be irrelevant. The day is nearly 24 hours long on Mars, but its year is nearly twice as long as ours. On the Moon, the year is just as long as it is here, but the day is two weeks long.

People living in various space environments may cling to a meaningless Earth calendar out of sentiment, or because it makes economic and cultural relations with Earth easier — but sooner or later there will be a movement for a rational scheme of measuring time, one that isn't tied to the movements of a single astronomical body. The rational scheme could be used for day-to-day living in space. What's more, a properly programmed computer can easily convert the rational scheme to the traditional scheme and vice versa.

What kind of rational scheme can we have? Here's one I suggest.

We could begin by keeping the day as it is, not because of its connection to Earth, but because the human being has evolved a natural cycle of rhythms with a period of about a day (circadian).

Instead of dividing the day by 24s and 60s, however, we would simply divide it after the usual decimal fashion into tenths, hundredths, thousandths, and ten-thousandths.

A hundredth of a day is 14.4 minutes long, or just about a quarter of an hour. Many people don't want the time closer than that for ordinary activities. If you ask "What time is it?" the answer "Point seventy-two" could be enough. It would mean the day was seventy-two one-hundredths over. If the day begins at midnight, it would be the equivalent of 5:17 P.M.

If you need a closer estimate of the time you could give it in thousandths or ten-thousandths. A thousandth of a day is 1.44 minutes, and a ten-thousandth of a day is just about 8.64 seconds. Giving the time to the ten-thousandth of a day ("It's point seventy-two thirty-four") is all you would need even if you didn't want to miss something as time sensitive as a spaceship takeoff.

People aren't in the habit of giving time as a pure decimal. There may be a transitional period in which they will feel more comfortable with names. Suppose you call a hundredth of a day a "nap" and a ten-thousandth of a day a "wink." If you're asked the time, you can say "23 naps, 42 winks," meaning that 0.2342 of the day has gone and it is what we would call 5:37 A.M.

Working in the other direction, 100 days might make a "season" (S) and 100 seasons a "generation" (G). Since 100 seasons (10,000 days) would equal about 27.4 Earth-years, that *is* about the length of a generation.

There would have to be some sort of general agreement as to the moment from which you would start counting the generations. Once that was settled, a day could be given as 15G, 84S, 06D, or once you took the generation-season-day system for granted it could be 15, 84, 06. In essence, we'd just be counting days, and this would be the 158,406th day, counting from some arbitrary first day. If you wanted to spot things to the nearest few seconds, it could be given as 15, 84, 06.0487.

It all sounds peculiar because we're not used to it, but how easy it would make bookkeeping and how convenient it would be in a thousand ways once we *were* used to it.

60 BY THE LIGHT OF THE EARTH

SOMEDAY, when the Moon is a busy world, with miners, electrical engineers, metallurgists, and astronomers all doing their work upon it, will tourists go there too? What would there be to do on the Moon? What would there be to see?

I suppose there will be tours through the lunar mining operations. People will watch the mass-drivers operate and will visit the great lunar telescopes on the far side. There might even be considerable fun in experiencing the low gravity.

My own feeling, however, is that the chief tourist attraction on the Moon will be the sky, for the fact is that the Moon has a much more magnificent sky than we have — when it can be watched.

You see, the Moon rotates very slowly with respect to the Sun, once every 29.5 Earth-days. That means it is two weeks between sunrise and sunset on the Moon and another two weeks between sunset and sunrise.

During the two-week-long period of daytime, tourists will have to stay in underground accommodations. The temperatures get too high, and the Sun's radiation, with no atmosphere to absorb any of it, is too dangerous. The two-week-long period of nighttime will represent the height of the tourist season for any given spot on the Moon.

For one thing, the stars will all be brighter and sharper than on Earth, since there will be no atmosphere to absorb any of the light or to make what light gets through waver and "twinkle." Many stars too dim to see from Earth will be visible from the Moon, and there will never be any

clouds. You will be in a planetarium that never turns off —
provided there are nothing *but* stars in the sky.

You see, the Moon doesn't rotate at all with respect to
the Earth. One side (the nearside, as seen from Earth) always faces the Earth; the other (the farside) always faces
away from the Earth.

On the farside, the Earth is never in the sky, and the stars
are the objects to watch. On the nearside, however, the
Earth is *always* in the sky, and no one is going to be watching the stars.

The Earth, as seen in the Moon's sky, goes through the
same phases in the same time as the Moon does in our
sky — but in reverse order. In other words, when we see
the full Moon on Earth, tourists on the Moon will be seeing "new Earth." On the other hand, when it is new Moon
on Earth, tourists on the Moon will be seeing "full Earth."

During the two-week nighttime (if we are standing at the
center of the nearside) Earth is at the zenith, and stays
there, except for a small wobble called "libration."

Earth is in the first quarter at sunset and looks like a
bright semicircle of light. The semicircle expands until, after
a week, we see the full Earth as a perfect circle of light.
Then the lighted portion of the Earth begins to contract
in the opposite direction until it is the last quarter (a semicircle facing the other way) at sunrise.

If you change your position on the nearside, the Earth
will shift its position lower in the sky. Once you have chosen
your new position, the Earth will also stay put in the sky.
Depending on your new position, full Earth may come
earlier or later in the course of the lunar night, but at
some time, during the course of that night, Earth *will* be
full.

And what an Earth it will then be. The Earth is larger
in the Moon's sky than the Moon is in ours; the Earth re-

flects a greater fraction of the light that reaches it from the Sun than the Moon does; and there is no atmosphere on the Moon to absorb the earthlight. Taking all these factors together, it turns out that the full Earth that shines on the Moon does so with no less than 70 times (!) as much light as the full Moon that shines on the Earth.

Imagine standing on the lunar surface in the bright earthlight. There would be enough light to read by, without uncomfortable heat or dangerous radiation. If you stand in the center of the nearside with full Earth at zenith, Earth's soft white light would cast virtually no shadow around you. If you stand elsewhere and the full Earth is lower in the sky, pitch-black shadows are cast, within which you can stand to gaze at the stars, if you wish.

Then, too, the Earth won't have the unchanging face of the Moon. It will be a blue and white circle of light, and through the curling banks of clouds you will see glimpses of brownish desert areas (especially if you use binoculars). What's more, the cloud layers will be forever changing as the Earth rotates and as its winds blow.

And yet the most beautiful sight of all will be in the daytime, after all, when, every once in a while, you *will* be able to emerge on the lunar surface and stare at the sky — as I'll explain in the next essay.

61 THE ORANGE CIRCLE

IN THE preceding essay, I discussed the appearance of the Earth as seen from the Moon's surface. In the Moon's sky the Earth would seem to stay in just about the same spot at all times. From a little more than half the Moon's surface, the Earth would always be visible somewhere in the sky, and where it is seen, there it would stay.

The Earth would be a spectacular sight, however, only when the Sun is not in the sky. Once the Sun rises, its brilliance and heat become steadily worse as it slowly climbs higher. (It is two full weeks between sunrise and sunset on the very slowly rotating surface of the Moon.) With the temperatures rising to the boiling point of water on those places on the Moon where the Sun is directly overhead, and with the Sun's ultraviolet light and x-rays unblocked by any atmosphere, the best place to be during the lunar day is well underground.

Even if you could imagine yourself on the surface in a protective suit of some kind, or were to watch on television, Earth would not be much of a sight by day. When the Sun is in the sky, the Earth is, on the average, farther from the full-Earth phase and therefore dimmer than when the Sun is not in the sky. Then too, the incomparable brilliance of the Sun, when it is in the sky, and the reflected gleam of the lunar landscape would wash out the brightness of the Earth.

But wait. As the Sun moves across the sky, it is bound to pass the Earth. Usually, the Sun will miss the Earth, passing somewhat above or below it. When the Sun passes it, the Earth is almost entirely black, for sunlight is shining on the other face of the Earth, the face turned away from the Moon. There is just the thinnest, faintest Earth crescent of light on the edge of the Earth facing the Sun, and it would, of course, be useless to try to see it with the Sun blazing away in the neighborhood.

However, the Sun's path changes somewhat from lunar day to lunar day, and every few passes across the sky, the

Sun moves *behind* the Earth. When that happens, the Earth's shadow falls on the Moon. (On Earth, we see this as a lunar eclipse.)

As viewed from the Moon, the Earth's eclipse of the Sun brings almost total darkness. The temperature drops precipitously, for there is no air to conserve the heat. During the eclipse it would be quite possible for people in the lunar base underground to come up to the surface in ordinary space suits to witness the sight.

That might not seem to be much of a sight, since the Earth would be totally black against a black sky. It would be invisible, wouldn't it?

So it would — except for its atmosphere. All around the black circle of the Earth is its atmosphere, and sunlight striking the far side of the Earth would also strike the atmosphere all around the edge and pass through.

Sunlight, passing through the thickness of the atmosphere, would be scattered. The short-wave light in the blue region would be particularly scattered and lost, while the long-wave light in the red and orange region would get through with comparatively little loss (as at the time of sunset on Earth).

All around the Earth, then, the atmosphere would light up with an orange sunset color. If the Sun passes behind the Earth off-center, and is closer to the top than the bottom (or vice versa), it will be a lopsided ring of light, brighter on the top than the bottom (or vice versa).

If the Sun passes behind the Earth dead-center, as it sometimes would, the ring of light would be very bright on the side the Sun entered and dim on the opposite side. As the Sun continued to move behind the Earth, the entering side would dim and the opposite side brighten until the orange ring (a perfect circle in appearance) would be equally bright everywhere. Then, slowly, it would be the opposite side that would brighten until, finally, the Sun emerges there and the spectacle is over.

Since the Earth is larger in appearance than the Moon, and since the Sun moves very slowly in the Moon's sky, the eclipse could last a couple of hours.

It is easy to imagine that the tourist season on the Moon would hit boom times when an eclipse is due, as everyone who can find rocket space crowds onto the Moon to stand on the surface under the orange light of the gigantic sunset that rings the Earth.

But there is a catch. Earth's clouds are sure to interfere with the brilliance and circular perfection of the orange ring. Parts of it will be dimmed or blotted out if the clouds are thick enough. On rare occasions, the cloud pattern on Earth might virtually block out all the ring.

I imagine that this would be considered an act of God and that no tourist would get his or her money back.

62 WHERE ALL IS STILL

ASTRONOMERS worry about a special kind of pollution — lightwaves and radio waves.

Through almost all of history, Earth's surface at night was dead black, with only an occasional forest fire making a small flicker. Human lights were too faint to matter.

Then, in 1877, the electric light was invented, and ever since, more and more each year, cities have grown brighter by night. The brightness has spread outward into the suburbs and along the highways, until it has become hard to find a place so isolated that the stray rays of man-made light do not flicker and blur the delicate tracery of the stars, nebulas, and galaxies of the sky.

What's more, since 1902, humanity has been producing radio waves in greater and greater variety and with stronger and stronger intensity, and these too fill the Earth's atmosphere with a "noise" that threatens to blot out the broadcasts of the distant galaxies.

Our telescopes, whether optical or radio, may become increasingly useless.

Space will give us a chance to get away from the interference of our atmosphere (see Essay 50, "Eye in a Vacuum") but, in orbit, there will always be times when the Sun, the Moon, or the Earth (or some combination) will be in the field of vision. This would be good if those bodies were being studied, but what if it were the rest of the universe?

Is there any place where we can be permanently out of sight of all three interfering bodies — a place close enough to be reachable? The answer is no — but almost.

Earlier I discussed the surface of the Moon, stressing the nearside that always faces the Earth as our satellite moves in its orbit.

Because the Moon's motion about the Earth is slightly uneven, about 60 percent of its surface has the Earth in the sky at least part of the time. That means, however, that 40 percent of its surface makes up the farside, where the Earth is *never* in the sky.

No light from the Earth, nor any radio waves of human origin, can penetrate through the thickness of the Moon. If one is on the point of the surface directly away from the Earth, there are 2160 insulating miles of lunar rock between that point and the Earth.

To be sure, there is still the Sun. It makes its existence felt on the far side of the Moon, but only intermittently.

When the Sun is present, it remains in the sky for two weeks between sunrise and sunset. But then it is absent for two weeks between sunset and sunrise.

For two solid weeks, the sky can be studied, with light-waves and radio waves from Earth, Moon, and Sun all absent. With no clouds, mist, or dust to interfere (since there is no lunar atmosphere), there will be a stillness, so to speak, that we can find nowhere else so close to Earth.

Telescopes could zero in on particular objects in the sky, any one of which would be moving across it from east to west at only 3.6 percent of the speed with which it would be moving in Earth's sky. As an example, variations in intensity of radiation from quasars, from galactic cores, from novas, even from the general microwave background, could be studied in detail, over comparatively long periods, with no interruptions.

When the Sun rises, of course, all that must stop. Indeed, astronomical instrumentation of all sorts would have to be withdrawn beneath the surface to prevent harm resulting from sudden temperature change or hard radiation — and remain withdrawn for two weeks.

Yet there would be special instruments to study the Sun during its two-week reign, when neither night nor drifting clouds nor even mist would dim its beauty. It will stay in the sky long enough to make half a revolution about its own axis so that just about every part of its surface could be studied in detail — over the entire spectrum — on each occasion that it passes across the sky.

It may be, then, that a large area on the side of the Moon directly away from the Earth would be set aside as an astronomical reservation, so to speak, with no other activity allowed. The astronomers themselves would have to restrict carefully their own use of light and radio.

One might almost imagine air-raid wardens patrolling the

area to make sure that no telltale flickers reveal the busy life going on beneath the surface or in the neighborhood of the instruments themselves.

63 WANTED: ORGANICS

THE BIG DISAPPOINTMENT of the Viking probes, which in 1976 landed on the surface of Mars and tested its soil, is that they found no organic compounds.

Organic compounds have molecules built up of chains and rings of carbon atoms to which hydrogen atoms (and usually other kinds of atoms as well) are attached. Organic compounds are the key components of all living organisms on Earth and, in fact, it is difficult to see how life (as we know it) can exist in their absence.

Before life came to be on Earth, there must have been a period of chemical evolution, perhaps hundreds of millions of years long, in which organic compounds existed and, under the lash of sunlight and other forms of energy, gradually grew more complex until very simple living things formed at last.

Biochemists have speculated on the course of such chemical evolution, but no traces of the period are left on Earth. They must rely on indirect experimental evidence based on what they think primordial Earth must have been like.

Much better evidence might be obtained if we could find some other world in its period of chemical evolution. The nature of the organic chemicals in the soil might yield key secrets concerning the evolution of life *on Earth*.

CHANGE!

Astronomers were quite certain the Moon was a dead world long before our astronauts reached it and brought back lunar rocks. There was some slim hope that organic compounds might exist in its soil, even if only in tiny quantities. They didn't.

Mars seemed a much better hope. It had a thin atmosphere that was almost all carbon dioxide. Since there are carbon atoms in the carbon dioxide, there might be carbon atoms in the soil too, as part of organic molecules. In that case, even if there was no life on Mars, it might be in the stage of chemical evolution.

But the Martian soil checked out negative too.

Where else can we go? Venus has a very dense atmosphere, 10,000 times as dense as that of Mars, also almost entirely carbon dioxide. Venus, however, is extremely hot, and organic compounds break down under the influence of heat. It is very unlikely, therefore, that organic compounds will be in the soil of Venus, or of Mercury either.

Yet there is tantalizing evidence that organic compounds *do* form when given a chance. Even if we don't count the organics here on Earth, we know that simple organic molecules exist in Jupiter's outermost layers, and more complicated ones may well exist. Unfortunately, Jupiter's actual structure is not something we can probe in the immediate future.

Surprisingly enough, simple organic molecules have been discovered in the vast dark clouds of dust and gas that lie between the stars in our Galaxy. They, however, are far more unreachable than Jupiter.

Are there no extraterrestrial organics closer than Jupiter, then? The answer is: Yes, there are! We don't even have to get off the Earth to find and study them, for they come here.

Small fragments of matter are constantly striking the

Earth. Most are the size of a pinhead or so, and friction with the air heats them white-hot so that they appear as meteors. Occasionally, larger bodies strike and survive the blazing passage through the atmosphere to reach the Earth as meteorites.

About 90 percent of the meteorites that fall and are salvaged are stony in composition. The remaining 10 percent are nickel-iron.

Every once in a while, though, a meteorite lands that is dark in color and crumbles easily. It is a rare carbonaceous chondrite and contains as much as 4 percent carbon. In 1969 such an object exploded over the town of Murchison, Australia, and 182 pounds of fragments were collected. In 1950 there was a smaller fall near Murray, Kentucky.

Such meteorites are found to contain small quantities of organic compounds. They contain molecules made up of carbon and hydrogen atoms (hydrocarbons) that resemble the fat molecules of living things. Also present are amino acids of the family of the simple building blocks that form the protein molecules characteristic of life.

The organic compounds in carbonaceous chondrites do not arise from living things, but are formed by natural processes that don't involve life. They could yield important information involving chemical evolution, if there were more such meteorites and if we could study them before they heated and exploded in their passage through the atmosphere.

Perhaps that can be done — as we shall see in the next essay.

64 THE DARK WORLDS

I POINTED OUT in the preceding essay that carbonaceous chondrites are in some ways the most interesting meteorites we find on Earth.

Iron meteorites (the most familiar) are made up of comparatively heavy atoms, stony meteorites of smaller ones, and carbonaceous chondrites of still smaller ones. Carbonaceous chondrites contain water together with carbon and organic compounds. The composition of carbonaceous chondrites might give us crucial information concerning the manner in which life evolves.

The trouble is that carbonaceous chondrites are very rare.

Yet in the universe as a whole, small atoms are more common than large ones. For that reason, stony meteorites are more common than iron meteorites, and one would think, on that basis, that carbonaceous chondrites would be most common of all.

Very likely they are, but the meteorites we find on the surface of the Earth have to survive speedy passage through dozens of miles of atmosphere, reaching white-hot temperatures in the process. Carbonaceous chondrites are composed of comparatively fragile materials that fragment and crumble easily, and very few survive the ordeal. Even those that don't melt and burn away into impalpable powders break up into small pieces by the time they reach the surface.

If we could study carbonaceous chondrites before they hit the atmosphere, if we could find them intact and unharmed, exactly as they were when they were formed four billion years ago, we could study them in detail. We could see how the light atoms combined and into which organic substances — something that must have taken place on Earth itself, four billion years ago.

But how do we find such objects, and where?

In the infancy of the solar system, the planets (and the Sun itself) formed by the accumulation and coming together of smaller fragments of matter. In general, most of the original cloud of fragments and particles collected into sizable worlds, but some remained behind as unfinished raw material, so to speak.

Most of the largest of these subplanetary fragments — the asteroids — whirl about the Sun in the space between the orbits of Mars and Jupiter. A scattering can be found outside those limits, and meteorites are thought to be examples of errant asteroidal fragments that have come to the end of their road by collision with Earth.

What are the asteroids made of? Until recently it was assumed that they were rocky worlds. Since 1970, however, sophisticated ways of measuring the way they reflect light show them to reflect surprisingly little — less than they would if they were rocky. This means that they are dark in color — as dark as the carbonaceous chondrites.

From this it is possible to estimate the density of some asteroids, a figure that turns out to be surprisingly low — as low as that of the carbonaceous chondrites. In fact, at least half of the asteroids may be, essentially, carbonaceous chondrites. That includes the largest of them all, Ceres, which is 620 miles in diameter.

At its closest, Ceres comes to about 160 million miles from Earth. Most of the large asteroids are at about that distance, give or take a couple of dozen million miles.

This is not an unreachable distance. We have probed Jupiter which, at its closest, is twice as far from us as Ceres, and Saturn, which is four times as distant as Ceres. Still, it would be nice if there were closer targets.

It may be that there are! One of the interesting pieces of information our recent Mars probes have discovered is that the two small Martian satellites, Phobos and Deimos, are quite different from Mars itself in surface appearance. The surface of Mars tends to be reddish and light, but the Martian satellites are dark — as dark as many of the asteroids. Astronomers are concluding that those satellites are, indeed, captured asteroids.

The Martian satellites would seem, in fact, to be carbonaceous chondrites. To be sure, they are tiny compared to Ceres. Deimos is, on the average, 8 miles wide, Phobos about 14. Yet they are enormously larger than any carbonaceous chondrite we are likely to find on the surface of the Earth — and for four billion years they have probably been untouched.

Well, we can reach them. We have reached Mars, but the two probes placed on it have tested the Martian surface for organic material and failed to find it. It might now be a very good idea to place a probe on the surface of Phobos instead, where it might find those organic compounds — and give us information about the beginning of the long climb that leads to life. (However, see also Essay 44.)

65 THE PUZZLING SUN

LIFE ON EARTH depends on the Sun. It not only gives us light and warmth, but green plants use the energy of its light to form the substances that serve as food for the animal world. Coal and oil are the products of ancient plants and animals that lived on energy from the Sun ages ago.

The uneven heating of the Earth by the Sun produces winds and ocean currents. The Sun's heat evaporates the ocean and makes rain and rivers (and hydroelectric plants) possible.

Yet with all that, it was only in the late 1930s that scientists managed to work out the source of all that essential energy. Hydrogen fusion goes on in the Sun's core, and scientists worked out, from laboratory data, the exact nuclear reactions that were taking place.

Some of those nuclear reactions give rise to neutrinos (see Essay 47), tiny particles that move at the speed of light and are affected only by the weak interaction, which will allow a neutrino to react with an atomic nucleus only when the two are very close — practically touching. Since a neutrino moves at the speed of light, it stays near an atomic nucleus for only about a trillionth of a trillionth of a second, and that's not long enough for anything to happen.

So neutrinos travel through matter as though it weren't there. Neutrinos produced in the Sun's core spray outward in all directions, passing right through the matter in the outer layers of the Sun. Those aimed in the right direction reach the Earth in eight minutes and then pass through the planet. In fact, there could be 200 trillion neutrinos from the Sun going through you every second without your ever being aware of it.

As long as neutrinos pass through matter without doing anything, they cannot be detected, and they exist in theory only. Every once in a while, though, one out of trillions of neutrinos might just happen to be snagged by an atomic nucleus as it slips past. When that happens, a change in the nucleus takes place, and particles that *can* be detected are emitted. In the 1950s, physicists worked out methods for actually detecting neutrinos in this way.

In the 1960s, efforts began for a systematic study of the

Sun's core through the detection and analysis of the neutrinos it emitted.

Raymond Davis, Jr., set up a vat of 100,000 gallons of perchloroethylene (a common solvent used in dry-cleaning solutions) in a gold mine in South Dakota. He placed it a mile deep, since a mile of rock would absorb any particles coming from space *except* neutrinos. Even the penetrating cosmic rays couldn't pass through a mile of rock, but neutrinos can.

Every once in a while an incoming neutrino (out of the many trillions passing through) would be absorbed by a chlorine atom in the dry-cleaning fluid, and that chlorine atom would be converted into a radioactive-atom variety of the gas argon. After you've waited long enough to allow a considerable quantity of argon atoms to accumulate, you flush out the system with helium gas and measure the quantity of argon that has been formed.

Davis began work in 1968, and he knew exactly how many argon atoms should be formed if theories about the Sun and our knowledge of neutrinos were correct.

In ten years of work, however, Davis never managed to collect more than one third of the neutrinos he was expecting. Sometimes he collected considerably less than one third.

This is puzzling and troublesome. Davis shouldn't be getting those results. Scientists don't know how to account for the missing neutrinos unless their theories about what is going on in the center of the Sun are wrong. And if scientists are wrong about the center of the Sun they don't know what other theory they can possibly work up to explain the facts.

The most frightening suggestion (but not a likely one at all) is that we have happened to catch the Sun at a time when it is undergoing changes, when it is beginning to work differently than it used to. It could be that such changes would eventually spell the end of life on Earth.

On the other hand, Davis's method only detects the most energetic neutrinos. Attempts are being made to devise other systems of neutrino detectors that will detect less energetic neutrinos and give us more details to work with.

The whole puzzle may, in a way, be a false alarm. In 1980 experiments were described that may indicate that neutrinos, which exist in three varieties, oscillate from one variety to another. The neutrino-detecting devices used to study the Sun are designed to discover only one variety — the kind the Sun emits. But suppose these neutrinos, en route to Earth, oscillate and distribute themselves among all three varieties, two of which pass the observers unnoticed. Naturally, they would spot only one third the number we would expect them to observe.

66 THE OCEAN WORLDS

WE USUALLY think of the Earth when we speak of an ocean world, for Earth's ocean covers 70 percent of its surface and is some two miles deep, on the average. The Moon, Mars, and Mercury, in contrast, have no ocean and, in fact, no free water of any kind.

A couple of decades ago, it was thought that Venus, considering its clouds, should have an ocean. There was speculation that the ocean might even be a global one, covering everything, leaving no land.

Wrong! It turned out that Venus was not much cooler

than red-hot, and there wasn't (and couldn't be) a speck of water anywhere on its surface.

But then our probes skimmed past the satellites of Jupiter and, behold, we found our ocean after all — a world-girdling one at that.

Four large satellites circle Jupiter, which, in order from the nearest to the farthest, are Io, Europa, Ganymede, and Callisto.

Io, the nearest, is about the size of our Moon. Being so close to Jupiter, it is strongly affected by the giant planet's tides, which tend to heat Io's interior and make it, literally, a world of fire and brimstone. Littered with active volcanoes that belch sulfur and sulfur compounds, it is covered with yellow orange lava that has filled any craters formed in the early days of the solar system. There is, of course, no water on hot Io.

The next satellite, Europa, was a complete surprise. It has neither craters nor volcanoes. It is, in fact, completely smooth and crisscrossed with long, straight lines. It is less affected by Jupiter's tidal effects, and never got warm enough to lose all its water. Plenty of water is still there, forming a global ocean — but a frozen one. The smooth surface we see is a glacier (with cracks) that covers the whole satellite. Europa is somewhat smaller than the Moon, so its ice-covered area is only twice as big as the ice-covered Arctic Ocean of Earth.

Europa's interior is probably warm enough to melt ice so there is likely to be a layer of liquid water underneath the glacier, which would explain the lack of craters. A meteor strike would break the ice and splash. The water, welling up, would freeze smoothly once again.

Europa's density is equal to that of our Moon, so it must have a rocky core, and the water may not be very deep.

The outermost large satellites, Ganymede and Callisto,

must be solid through and through, for each is littered with craters. Their density is distinctly lower than that of rock, so they must be made up of half ice and half rock.

Ganymede and Callisto are large satellites, considerably larger than our Moon, and the icy surface of each is equal in area to our Atlantic Ocean. Then too, the frozen oceans of those two worlds are deep indeed, for they may extend nearly to the center, without much in the way of liquid water in the depths, and perhaps none at all.

Someday, the ocean worlds that we have now observed close-up for the first time may become vitally important.

When human space settlements have crowded the lunar orbit uncomfortably, and when, a century from now, settlements are being constructed in the asteroid belt, resources will be needed there.

The asteroids themselves will serve as a supply for all the metals, for concrete, soil, glass, and even oxygen. To a limited extent, they will also supply the crucial light elements of hydrogen, carbon, nitrogen, sulfur, and phosphorus.

The light elements will be present in quantity, however, on the three ocean worlds of Europa, Ganymede, and Callisto, where there is a millions-of-years supply of water, probably carbonated, and undoubtedly bearing all the commoner mineral salts in solution.

Europa and Ganymede are located in Jupiter's giant and deadly radiation-loaded magnetic field, but Callisto is outside the worst of the field and is the most easily accessible of the three. A century from now, moreover, it will probably not be necessary for human beings to construct the settlements themselves. Automated machinery — deft, versatile, and all supervised from surveillance stations on Jupiter's asteroid-sized outermost satellites — will take care of things.

Blocks of crude ice will be mined and hurled into space, where they will be distilled and separated into pure water, frozen carbon dioxide, and a variety of minerals, all for the use of the asteroid settlers who will flourish, thanks to the resources of their distant Callistan ocean.

67 THE FUTURE MARINERS

REACHING THE MOON has been deceptively easy. It's only three days away. We can go there, explore for a day, and come back — all in a week. The only feasible method for human astronauts to reach any other sizable body beyond the Moon at present is to build up a speed of a dozen or so miles a second, and coast. Such voyages will take years.

The next reasonable target beyond the Moon is Mars, and the round trip by present techniques will take about two years. That might be carried out as a tour de force, but surely it will be difficult for human beings to commit themselves for two years to the inevitably cramped quarters of a spaceship as large as we are likely to build in the next couple of decades — with the only relief coming from the leg-stretching stay on Mars itself.

Even if that could be done, what about the exploration of the solar system beyond Mars in voyages of, inevitably, much longer duration?

But why Earthpeople for these trips anyway? We are unfitted, psychologically, for the exploration of space. Suppose, instead, that we continue the unmanned probing of the other planets, while we concentrate our manned efforts on exploring and colonizing the Moon and on building the kinds of space colonies in the Moon's orbit that I described in Essay 53.

Once the Moon is colonized and self-supporting, or space

colonies are well-organized (or both), we could safely leave it to the colonists to carry on space exploration. It is *they* who would be the future mariners, the explorers who would plumb greater emptinesses than were ever dared by those ancient mariners, the Phoenicians, Vikings, and Polynesians.

Why should this be? Consider —

1. Space flight is an exotic matter to Earthpeople, something not of the fabric of our life, that would take us away from our age-old world.

The colonists, on the other hand, would make their lives on a world that they or their ancestors had only reached by spaceflight, and they would always be involved in continuing spaceflight for reasons of commerce and travel. Spaceflight would be second nature to the colonists.

2. The conditions of spaceflight would represent an extreme turnabout to us of Earth. We are accustomed to clinging to the outer surface of a very large world; to a cycling of food, air, and water through so large a system, we are scarcely aware of the process; to a complex ecology, copious water, extensive atmosphere, endless room. To change to a spaceship where we must live inside narrow walls, count every resource, cycle everything so tightly that we are constantly aware that our food and water is derived from wastes, would be to give us an unbearable feeling of constriction and claustrophobia.

The colonist, however, is used to exactly that. He lives in tunnels on the Moon or in cylinders out in space, under spaceshiplike conditions. The spaceship is, at worst, somewhat narrower than his world, but he is inured to its disadvantages.

3. Earthpeople are used to a constant and strong gravitational pull. Even if a spaceship is made to spin on an axis to produce a centrifugal effect that mimics a gravitational pull, it may be that the pull will be less than we are used to in order to avoid the disadvantages of a too-rapid spin. Or

if the effect is that of a full Earth gravity, that force will nevertheless diminish as one moves toward the axis of spin. Earthpeople may not be able to adapt to low or variable gravity easily.

Colonists would, however, have become inured to less than normal gravity on the Moon (where the force is only one sixth that of the Earth) or to variable gravity on the spinning space colonies. They would have been born to it, and spaceship gravitational effects would not discomfort them.

4. The worlds human beings are likely to reach and land on in the outer solar system — the asteroids and the satellites of the giant planets — are much more Moonlike than they are Earthlike. They would be oppressive and depressing to Earthpeople, who, landing on them, would not find in the solar system any place that reminds them of home.

The colonists would not feel this psychological break. They could tunnel into Ceres or Callisto as they had tunneled into the Moon. They could build space colonies in the asteroid belt as they had built them in the Moon's orbit. The new neighborhood would seem fine to them.

The conclusion is that human beings will surely explore space properly, but only after they first colonize the space in Earth's vicinity.

68 BELOW THE SPEED LIMIT

HOW FAST can we go?

We are told over and over again that the speed of light

in a vacuum is the absolute cosmic speed limit — 186,282.4 miles per second, and that's it. There are cosmic ray particles that go at 99+ percent of the speed of light, but nothing with mass goes faster.

What's the best way of approaching that speed limit?

Well, the energy-gobbling chemical-rocket engine is suitable for getting into space, but once there we might switch to an ion-drive engine, which emits speeding ions (electrically charged atom fragments). The ions move much faster than the gases in an ordinary rocket exhaust, so the ions give a much more efficient kick forward to the spaceship. Of course, the ion is submicroscopic in size, so the kick forward is submicroscopic in size too. Even if billions of ions shoot backward each second, the spaceship accelerates very slowly — but steadily.

After the ship has been in space a year under this tiny, but constant, acceleration, it would indeed be moving at speeds approaching those of light. Though it is impractical to try to carry enough chemical fuel for that much acceleration, it *is* practical to carry enough ion-yielding material to do it. The ion-drive loses you time, but it gains you portability.

Even the speed of light, however, is a crawl in comparison with the size of the universe. It would take a hundred thousand years (to those of us on Earth watching the progress of the journey) for a ship traveling at the speed of light to cross our own Galaxy. It would take twelve billion years to reach the farthest object we can see in our telescopes.

Yet, though the speed of light is rendered so insignificant by the size of the universe, it may be that we won't even be able to do *that* well.

After all, a railroad train might very easily be able to do 120 miles an hour, but the condition of the roadbed may keep it down to 70 miles an hour. An automobile capable of cruising at 80 miles an hour may be forced down to 20 miles an hour by fog.

What's the condition of the roadbed in space? How's the fog?

The stars can be discounted. They surround us by the tens of billions, but the space they're spread through is so enormous that their numbers are not important. Stars are so far apart that if a spaceship were to travel in a straight line through the Galaxy, at random, the chance that it will come anywhere near even a single star is very small.

Who says, however, that stars are the only objects in space? There must be smaller bodies as well, bodies too small to develop into stars and give off light. That means they are too small to be detected at a distance. They would be detected only when they were so close to the ship that, traveling at the speed of light, that ship would have no time to avoid a collision.

A glancing collision of a ship traveling at nearly the speed of light with even a small asteroid would be instantly fatal to the ship and to everything aboard.

What are the chances of such a collision? We can't say, for we don't really know how thickly strewn with chunks of debris the space between the stars is. Interstellar space may be surprisingly clear of such objects. Would we *then* be safe in moving at near the speed of light?

No, for we *do* know that interstellar space contains dust particles, as well as individual molecules and atoms. We can actually see dust clouds and detect some of the atoms and molecules within it. Even space that is otherwise clear contains a thin scattering of hydrogen atoms. The clearest area is probably that between galaxies, and it is estimated that in space there is one hydrogen atom every cubic yard.

Hydrogen atoms matter. If a ship, going at nearly the speed of light, strikes a hydrogen atom, that is equivalent to a hydrogen atom, going at nearly the speed of light, striking a ship. Such a speeding hydrogen atom is a cosmic ray. If there were only one hydrogen atom every cubic yard, a spaceship going at nearly the speed of light would be exposed to something approaching a billion billion cosmic ray particles every second. The ship would quickly become radioactive, and everyone on board would be fried.

That's the most favorable condition. Inside a galaxy there would be perhaps a million times as many strikes, and if one passed through a gas cloud...

Barring unforeseen breakthroughs, then, it may never be practical to travel through space at more than one tenth the speed of light — if that.

---

69 TO THE STARS!

IN ESSAY 67, I talked about the exploration of the solar system by human beings who had colonized the Moon or lived in artificial worlds built in the Moon's orbit. Even their journeys of years and decades would, however, explore only the outer solar system.

Is this enough? The solar system, very possibly without life, is our backyard, already known in part.

What about the stars? There are over a hundred billion of them in our Galaxy alone, and there must surely be hundreds of millions of Earthlike planets circling other stars. Such planets may bear life, possibly even intelligent life. Will the day come when we can explore other planetary systems?

The great difficulty is distance. It takes light 4.3 years to reach the nearest star, Alpha Centauri, and nothing can go faster than light in our universe. It would take light dozens of years — hundreds of years — to reach more distant stars, and a hundred thousand years to cross our Galaxy.

So far, our most rapid rockets can only move at about a ten thousandth of the speed of light. Coasting at that rate, it would take 43,000 years to reach Alpha Centauri.

Is it possible, perhaps, to make use of space drives that could carry us easily at far greater speeds than chemical rockets can attain? In particular, might we find a way, some-day, to move faster than that ultimate limit, the speed of light?

Scientists have speculated about the existence of a universe of tachyons, particles that go *only* faster than light. Could we turn a ship and its crew into tachyons, have it speed off and reach the neighborhood of Alpha Centauri in mere hours, then turn it back into ordinary particles?

The trouble is that some scientists doubt the possibility of tachyons, and certainly they haven't yet been detected. Then too, even if they existed and were detected, we don't have the faintest idea how to convert ordinary particles into tachyons, or vice versa; or how to control a tachyon ship if we could form one. This is not something firm enough to build dreams on.

We might simply use new space drives to build up speeds close to that of light. The theory of relativity predicts that the astronauts on board the speeding ship would experience a slowdown in time, so it would *seem* to them they had spent only a few months, perhaps, in reaching a distant star.

Attaining such speeds in any fashion, however, is likely to use more energy than can reasonably be packed into a ship, or gathered in space en route. Such speeds may also produce too much friction with the scattered atoms and dust particles in interstellar space. Besides, though the astronauts might experience slow time, back in the solar system (still on ordinary time) decades or even centuries would pass. The astronauts might not be willing to volunteer for a trip that would take them away forever from the world they knew.

The same argument can be used against long journeys undertaken at low speed, by astronauts who would be frozen and therefore not experience the passage of time. We don't know whether we can freeze human beings so that they can be resuscitated safely after decades or centuries. Even if we did, the astronauts would still have left forever the worlds they knew.

But suppose we abandon the untried technologies involved in superrapid speeds and life-freezing. Suppose we stick to low speeds and full consciousness, which we know about. Imagine the assembly of a huge starship in space that can be sent coasting to distant stars without problems of energy or friction. Suppose it is large enough to hold tens of thousands of men, women, and children, to grow its own food, recycle its own wastes, and be a self-contained world.

The astronauts of both sexes on board such a ship would have to spend the rest of their lives on it, as would their children and their children's children. Many generations might have to endure the voyage before reaching an interesting star.

Is this likely? Would people be willing to pen themselves within a ship — and their children and grandchildren too?

Not Earthmen, perhaps. If, however, we build space colonies in the Moon's orbit (see Essay 53), those space colonies would be exactly the kind of starship I have just described.

The people on such a space colony may someday decide

it is no longer interesting to go round and round and round the Earth and Sun. Using some efficient space drive based on fusion energy, they may decide to fly straight and move off into illimitable space.

There'll be no expense, for the starship would already exist. There would be no trauma of leaving home, for home would be coming along. And through space colonies the whole universe may someday open up to humankind.

70 QUARANTINE

LET'S CONTINUE considering the possibility that someday space settlements might choose to break free of the solar system, move away from the Sun and its planets, and drift outward toward the stars.

But why should they do this? One might advance such motivations as curiosity, or the pioneering spirit, but is that very convincing?

A much better motivation has now occurred to me, and, to explain, I must begin with Earth.

In some ways, Earth represents a single ecosystem with only one atmosphere and one ocean. Birds can range widely through one, and sea creatures through the other, while land animals and plants can spread through whole continents.

Nevertheless, there is also the chance for isolation. New Zealand was isolated by surrounding ocean for so long that land mammals, evolving elsewhere, never had a chance to reach the New Zealand islands, which made them a natural home for a variety of large flightless birds not found else-where. Australia, isolated for a lesser time, did have land mammals, but only of the primitive monotreme (egg-laying) and marsupial (pouch-bearing) varieties. Small isolated is-lands scattered over the ocean, hidden valleys and canyons, and inaccessible forests can be havens for species not found elsewhere.

These isolated paradises are not always good for the in-habitants. Security prevents the proper tempering in the fires of competition. When the outside world creeps in with more competitive animals, then good-bye inhabitants.

South America was once an island with a wide variety of large, primitive mammals. When the isthmus of Panama was formed, and more advanced mammals moved down-ward from North America, most of the South American mammals were wiped out.

Human beings have reached all the land areas of the world, bringing with them their animals. The native species of the lands penetrated, unable to cope with the invasion, were destroyed by the united action of human hunters and animal predators.

Animals can seize and take over new territories when human beings deliberately or accidentally carry them from one place to another. Starlings have colonized North Amer-ica, and rabbits have colonized Australia against the will and to the despair of human beings. Germs travel with their human hosts, and epidemics can be worldwide.

How will such things affect space settlers?

Each space settlement may well have its own ecosystem, each much simpler than Earth's. The space settlers on each little world will undoubtedly try to exclude weeds, vermin, and disease germs to an extent consistent with a balanced ecology. It could be that each different space settlement would have a somewhat different ecosystem.

In that case, how will trade be carried on? How will human beings from different settlements work together on projects in space? Here on Earth, we have regulations forbidding the importation of plants and animals — quarantine regulations to prevent the spread of disease. Much tighter rules would be necessary in space settlements!

Everything entering one space settlement from another would have to be closely scrutinized to prevent invasions of unwanted life forms. And if one space settlement suspects another, all will unite in suspecting Earth. After all, Earth will have the largest, most unruly and chaotic, and most dangerous ecosystem known. It would be the reservoir for every infectious organism, every noxious parasite, every undesirable species.

Some settlers might be resigned to the situation and realize that the ecosystem on a settlement cannot remain isolated. They will mix with other settlements and, in the end, might have a single space ecosystem with variations from settlement to settlement, but without absolute differences. Even to these assimilationists, however, Earth itself would cause great anxiety.

Still other settlers could withdraw into biologic isolation. They could argue that it is better to cut down on trade and cooperation rather than risk the pollution of what they might consider an ideal ecosystem.

And in the end, in search of the ultimate quarantine, they could leave the solar systems, taking themselves and their plants and animals off into the purity of distant space, where nothing but the slow forces of evolution could change the balance they had set up.

That might be the motivation for action that could lead to the seeding of the universe and the colonization by human beings of other planetary systems and of space itself.

It would not be the pull of distant worlds and the call of great spaces. It would be their fear of the worlds in which they live and the horror of pollution.

71 THE COSMIC SUBWAY LINE

THE MOST exciting phenomenon in astronomy these days is the black hole — an apparent final graveyard of matter, thanks to its gravitational field.

Only four kinds of forces are known to exist in the universe, and gravity is by far the weakest of the four. But wait.

Two of the forces are very short-distance phenomena that involve only subatomic particles and aren't ordinarily felt outside atomic nuclei. A third one, electromagnetism, is long-distance, but expresses itself attractively under some conditions and repulsively under others. The two tend to cancel each other, so that electromagnetism never manages to display really great intensity.

Gravity is different; it shows itself as a long-distance phenomenon, and only attractively. The more matter you pile together in one place, the greater its gravitational field becomes. If you start with a certain amount of matter and squeeze it together more and more tightly, the stronger its gravitational field becomes. Either way (or in combination) a gravitational field can be made greater than any other possible force.

As gravitation becomes extreme, all matter within its influence breaks down. Atoms and even subatomic particles squeeze down to nothing. Anything that falls into a sufficiently intense gravitational field can't ever come out at the point it entered, so that the field acts as a hole. Even light can't emerge, so it is a black hole.

A black hole can form when a large star explodes and collapses. Astronomers think that an object they call Cyg X-1 is a large black hole in our own Galaxy. Black holes of all sizes may be distributed all over the universe. Even mini–black holes, no more than pinpoint sized, may have formed in the great explosion that produced our universe in the first place.

Perhaps all matter will eventually fall into one black hole or another, until only black holes are left, and that would represent the final end of the universe. Such an end is many billions, perhaps trillions, of years in the future, however. Meanwhile, black holes could conceivably be put to use.

Objects spiraling into black holes gain vast energies of motion from the black holes' gravitational field, and some of this energy is converted into intense radiation. An advanced civilization (perhaps we ourselves, someday) may set up outposts near a black hole — but not too near, of course — to tap this overflow of energy.

We might even imagine methods devised to force wandering objects closer to the black hole, close enough to push them into the ultimate inward spiral from which they will never return, and in the course of which floods of usable energy would be emitted, absorbed, and stored. The black hole would thus be treated as a huge furnace for which any sort of matter would serve as fuel.

But what happens to matter that enters a black hole? Some astronomers think it isn't really lost forever, but is extruded, like toothpaste, into another part of the universe. At the point of emergence it would expand and blaze with energy as a white hole. Perhaps the mysterious quasars, far-distant objects that gleam with the light of a hundred galaxies at once, are white holes.

Under the extreme conditions of the black hole, matter may travel from one place to another far-distant place in very little time, transcending the speed limitations of the ordinary universe.

The Cornell astronomer, Carl Sagan, wonders whether the day might not come when mankind would learn enough about black holes to devise methods for surviving the conditions within them. Perhaps special gravity-resistant ships, using scientific principles undreamed of today, could carry men and goods through black holes to that distant terminus at the other end.

There may be many black holes (Sagan estimates perhaps as many as a billion in each galaxy, with the average distance between them 40 light-years — just a hop and a jump on the cosmic scale), each representing a different route, going from some particular place to some other particular place. Little by little, humankind might be able to map out the routes of these cosmic subway lines and work out schemes for traveling from any point in the universe to any other point by some appropriate combination of black holes.

Or perhaps some other, more advanced, intelligence (or groups of intelligences) in the universe has already succeeded in doing this. Perhaps a Cosmic Empire exists, with prosperous industrial planets located not too far from some black hole terminus. It may be that we won't have to map the universe at all, but that, when the time comes and we are sufficiently advanced, we may simply join the Cosmic Empire and become full members of the universe at once.

If it is true that black holes represent not the death of matter, but its death and resurrection, the universe would last forever. And with it, the various intelligences, including humankind's, might last forever as well.